SMART METHODS FOR ENVIRONMENTAL EXTERNALITIES

Smart Methods for Environmental Externalities

Urban Planning, Environmental Health and Hygiene in the Netherlands

GERT DE ROO, JELGER VISSER and CHRISTIAN ZUIDEMA
University of Groningen, The Netherlands

ASHGATE

Published by
Ashgate Publishing Limited
Wey Court East
Union Road
Farnham
Surrey, GU9 7PT
England

Ashgate Publishing Company
Suite 420
101 Cherry Street
Burlington
VT 05401-4405
USA

www.ashgate.com

British Library Cataloguing in Publication Data
Roo, Gert de.
Smart methods for environmental externalities: urban planning, environmental health and hygiene in the Netherlands. – (Urban planning and environment)
1. City planning – Environmental aspects – Netherlands. 2. Environmental policy – Netherlands.
3. Environmental Health – Planning – Netherlands.
I. Title II. Series III. Visser, Jelger. IV. Zuidema, Christian.
307.1'216'09492-dc22

Library of Congress Cataloging-in-Publication Data
Roo, Gert de.
Smart methods for environmental externalities : urban planning, environmental health and hygiene in the Netherlands / by Gert de Roo, Jelger Visser, Christian Zuidema.
p. cm. – (Urban planning and environment)
Includes bibliographical references and index.
ISBN 978-1-4094-2544-1 (hardback : alk. paper) – ISBN 978-1-4094-2545-8 (ebook)
1. City planning–Environmental aspects–Netherlands. 2.Environmental policy–Netherlands.
3. Environmental planning–Netherlands.
I. Visser, Jelger. II. Zuidema, Christian. III. Title.
HT169.N4R665 2011
307.1'21609492–dc23

2011023630

ISBN 9781409425441 (hbk)
ISBN 9781409425458 (ebk)

MIX
Paper from responsible sources
FSC® C013056
FSC www.fsc.org

Printed and bound in Great Britain by
TJ International Ltd, Padstow, Cornwall.

Contents

PART C RELATIONS AND REFLECTIONS

List of Figures

List of Tables

Preface

Smart Methods for Environmental Externalities is about methods that have been created in the Netherlands. This is rather special and unusual for an international publication on environmental policy and spatial planning. It is felt that the Dutch methods for environmental externalities have evolved in a rather unique way that is incomparable to any other country in Europe and beyond. The evolution of the methods has resulted in a wide range of alternatives for policymakers and government officials to choose from when considering issues such as environmental health and hygiene, spatial development, urban renewal, neighbourhood renovation and the like. The wide range of methods for environmental externalities can therefore be of interest to an international audience, dealing with environmental intrusion and the liveability of the urban environment. We would therefore like to express our sincere thanks to our peers and our publisher for embracing this argument.

In the past we have presented Dutch methods dealing spatially with environmental externalities to the international arena, in various publications, but not yet in a comparative and coherent form. We have published books categorizing methods for environmental externalities, but not for an international readership. This book, *Smart Methods for Environmental Externalities*, addresses the unique evolution of Dutch initiatives, which has resulted in a wide range of complementary methods. For us, this meant putting aside those methods, such as Environmental Impact Assessments, that are known, accepted and used globally, despite the fact that EIS is used in the Netherlands with a Dutch 'twist'. Neither have we discussed in great detail the international developments regarding methods for environmental externalities, despite the fact that we know they exist. It would have been a challenge, but a different form of research, that is beyond the scope of this particular book. You will be able to read more on that subject in other books in the Ashgate series *Urban Planning and Environment*.

Although there are several methods – in the Netherlands and internationally – for dealing with environmentally intrusive and environmentally sensitive functions, they do not have a general name. The Dutch language does not offer much help in this respect. As we have been researching environmental-spatial conflicts for many years, we began presenting the methods for dealing with these conflicts as 'environmental-spatial methods'. Other names have also been proposed: 'methods for environmental awareness' and 'area-specific environmental assessment tools'. No doubt more obscure proposals have also been suggested over time. Eventually the name was narrowed down to 'methods dealing with the spatial consequences of environmental externalities', which we quickly shortened to 'spatial methods for dealing with environmental externalities' and subsequently, for practical and publicity reasons, further abbreviated this to 'methods for environmental externalities'.

We believe that the methods are just that, namely 'for environmental externalities', with 'externalities' intrinsically addressing the spatial perspective of issues. The addition of the word 'Smart' to the title of this book was not just for the sake of publicity. We could explain S.M.A.R.T. as referring to Self-Monitoring, Analysis, and Reporting Technology, and no doubt a number of the methods presented would meet this criterion. But that would be a retrospective and 'super-imposed' construction. For us, 'smart methods for environmental externalities' refers to the innovative and integrative qualities of the methods presented, bridging two policy disciplines, namely environmental policy and spatial planning.

It has taken far longer than usual to bring this book to publication. Given that it takes, on average, a couple of years to bring out a book, one can imagine how long this particular process has taken. The process began with a request from the Dutch Ministry of Housing, Spatial Planning & the Environment (now part of the Ministry of Infrastructure & Environment). The report that was compiled in response to the Ministry's request was considered by all parties at the time to be of sufficient interest to warrant publication for a wider audience. We are extremely grateful to the Ministry for supporting us in rewriting the text of the report for publication as a book, and in having it translated into English. In particular we would like to thank Jan Jaap de Boer, Peter Kiela and Hans Verspoor, all from the Ministry, who kept faith in the project over the years.

One reason this all took so long was that Jelger Visser, one of the authors, fell ill and was unable to continue participating in the project. It must be quite a surprise for him to see the book completed at last. We thank him immensely for leading the project during the first couple of years, and we sincerely hope we have done what he expected of us to complete the book. A crucial link in the chain has been Yvette Mead of the University of Groningen Translation and Correction Department, to whom we could send bits and pieces which she converted without hesitation into proper English for us. Thanks Yvette, for all you've done through all those years. Our thanks go also to Koen Klieverik and Rens Baltus for providing logistic support during the project. Professor Donald Miller, the series editor, deserves a 'thank you' for embracing this work and accepting it as part of the 'Urban Planning and Environment' series. Last but not least, very warm thanks go to our publisher at Ashgate, Valerie Rose, who somehow has unshakeable confidence in whatever we propose. We do hope this project will justify that confidence too.

Gert de Roo
Christian Zuidema
Groningen, September 2011

Chapter 1
Smart Methods: Methods in a Changing Environmental Policy Climate

Smart Methods for Environmental Externalities is a book totally dedicated to innovative initiatives generated in the environmental policy and spatial planning arena in the Netherlands. Over the years, and since approximately 1985, various methods have emerged relating to the effects of environmental externalities on urban development. The result is a rich and wide range of methods that take into account a variety of aspects relating to environmental externalities. On the one hand, these wide-ranging methods together constitute a wonderful toolbox for policymakers, spatial developers and government officials, but on the other hand it is no longer obvious which method to choose. Therefore, this book not only presents the variety of Dutch methods for environmental externalities but also proposes a categorization for the methods that is clear and coherent, and will be considered useful in selecting methods for a specific issue.

The categories for methods for environmental externalities are presented in Part B of this book. The logic behind the categories requires a brief explanation. The Dutch methods for environmental externalities are, to a certain extent, a direct response to policy practices. The methods are tools for making policy practices visible and workable, and for translating environmental policy directives into spatial consequences. The methods for environmental externalities therefore reflect the changes that have taken place in environmental and spatial policies.

In the 1980s, environmental policy directives were driven by top-down, quantitative standards, constraining spatial developments in order to guarantee a healthy local environment. Environmental standards were used to physically separate environmentally intrusive and environmentally sensitive functions, thereby ensuring a safe distance between residential areas and industrial sites. The same applies to housing versus traffic and infrastructure. The first methods for environmental externalities therefore focus strongly on setting conditions, building on environmental standards and presenting their spatial consequences.

In the early 1990s, almost 'hand in hand' with planning theory – which underwent a communicative turn in response to a failing technical-rational approach to planning – the focus of environmental policy in the Netherlands shifted from top-down and technically driven policy to policy that took account of local conditions. This shift resulted in the decline of the environmental standard in Dutch policymaking. We see instead the rise of communicative approaches, area-specific and tailor-made policymaking and the decentralization of responsibilities and initiatives. With this shift, environmental issues no longer constrain spatial developments but should contribute to more sophisticated approaches, integrating the spatial, infrastructural, social and environmental aspects of a specific location.

The methods for environmental externalities have evolved accordingly. The evolutionary path will be elaborated upon in Part A of this book. While Dutch environmental policy is no longer the inspiration to the world that it once was, and in the light of the considerable uncertainty as to how Dutch spatial policy should move forward, the Dutch methods for environmental externalities still stand, and are used and tested under various conditions. The result is a range of methods that, together, constitute an excellent toolkit from which the 'right' method can be selected to tackle a situation involving environmental externalities, whatever that situation might be. There are methods for dealing with straightforward, relatively simple environmental-spatial conflicts, methods for dealing with complex environmental-spatial conflicts, and methods that can be used in the most chaotic of conflicts that are overly complex due to the opposing interests of numerous stakeholders and to the intangible 'jumble' of intrusive and sensitive functions in the urban environment. It is this range of methods and its internal coherence that are reflected in the title of the book, *Smart Methods for Environmental Externalities*.

1.1 Prologue: An Emerging Policy Field[1]

In 2001, the Dutch cabinet presented the policy document 'Where There's a Will There's a World. Working on Sustainability'. This document, the fourth National Environmental Policy Plan (NEPP-4), deals with more than environmental policy for the coming years. It also assesses thirty years of environmental policy in the Netherlands. During that time, environmental policy was developed from scratch. The message is clear: a great deal has been achieved, but there is still a long way to go. NEPP-4 identifies seven persistent environmental problems that the policy has not yet come to grips with. These vary from global climate change to negative impact on the quality of the living environment. In order to solve the last problem, NEPP-4 proposes a reform of policy on the living environment. This policy reform means that other levels of government 'will be afforded greater freedom and as much integrated responsibility for the local living environment as possible, including the related instruments' (VROM 2001; 329).

An important step in that direction was taken on 13 May 2004. The MILO method was presented at the conference 'Kwaliteit van de leefomgeving' (Quality of the Living Environment).[2] The audience included environmental officials and spatial planners from municipal and provincial authorities, local and regional

1 This section is a survey of the development of environmental policy in the Netherlands. Chapter 2 deals in more detail with the different periods in environmental policy and their significance for methods for environmental externalities.

2 MILO is the Dutch acronym for 'Environmental Quality in the Living Environment'. The MILO method is a joint project of the Ministry of Housing, Spatial Planning and

environmental agencies and water boards. MILO is a practical method that will enable them to improve liveability and the quality of the environment. It is definitely not a blueprint but 'a tool and source of inspiration for policy practice' (VNG et al. 2004; 5). There was a good reason why this policy practice was the source of inspiration for the tool: rather than following national programmes, policy proposals are shaped by local circumstances. MILO builds on practical experiences with methods – referred to in this book as methods for environmental externalities – that can be used to streamline the harmonization of the environment and spatial planning.

MILO is certainly not the only method developed to support local and regional environmental policy and integrated environmental policy. Over the years, various organizations in the Netherlands[3] have taken the initiative to develop their own methods, so that the range now available is extensive and above all diverse. As a result, the Netherlands has established a unique position in the world, both in terms of environmental policy and methods for environmental externalities.[4] This book is about spatial methods regarding environmental health and hygiene and is intended to serve as a guide to selecting a suitable method for aligning planning activities and the quality of the living environment.

MILO is clearly a product of today. The 'area-specific approach', 'quality ambitions' and 'area types': these are all MILO concepts that were not yet used in the early years of Dutch environmental policy. After a period of social and political awareness, in which the publications *Silent Spring* by Rachel Carson (1962), the renowned report by the Club of Rome (1972), and the UN environmental conference in 1972 played a role, the environment was suddenly high on the agenda. In the early period, this was reflected above all in the cleaning-up of the largest and most serious forms of environmental pollution. In Urgency Policy Document on the Environment[5] of 1972 – seen by many as the beginning of environmental policy

---

the Environment (VROM), the Association of Netherlands Municipalities (VNG), the Interprovincial Council (IPO) and the Association of Water Boards (UVW).

3 The Netherlands Ministry of Housing, Spatial Planning and the Environment is not the only body actively involved in developing this type of method. Local and regional authorities and environmental services have also developed several methods. A detailed overview can be found in the *Kennisboek Milieu in stedelijke vernieuwing* (VROM 2002) and on the SenterNovem website (Senternovem 2007).

4 Developments in Dutch spatial and environmental policy are followed with interest in various parts of the world. In Japan there is even a Study Group for Dutch Spatial Planning (De Roo 2002; 16). The Netherlands is apparently an example for other countries to follow. This role is reflected in the transformation of a Dutch environment-aware method in the United States. This will be discussed later in the chapter.

5 The document is characterized by the fact that it dealt only with environmental nuisance and environmental contamination. In particular, the Urgency Programme in the policy document reveals the close relationship to public health: environmental problems are only regarded as such if they constitute a risk or potential risk to public health.

in the Netherlands – the Biesheuvel cabinet still believed that this clean-up policy would take only five to ten years. In practice, however, this proved too optimistic.[6]

In the early years of environmental policy in the Netherlands, the belief in the 'makeability' of society was still evident.[7] The prevailing line of thinking was that tightly coordinated government intervention could fundamentally improve society. It is also one of the ideas that shaped the early development of environmental policy. It is reflected in the Urgency Programme of the Urgency Policy Document on the Environment: 'In the coming period, priority will be given to extending statutory measures'[8] (VM 1972; 23). The publication of the policy document on ambient environmental standards (*Nota milieuhygiënische normen*) laid the foundation for the current interpretation of the concept of 'environmental quality'. Generic environmental standards were seen as the instrument that would give shape to environmental policy, and the mechanisms of the first methods for environmental externalities are largely based on this concept of quality.

Under the later – more moderate – cabinets of Van Agt, belief in the makeability of society began to erode. Results were not achieved as easily as was assumed in the Urgency policy document. Society is only makeable to a certain extent. As the belief in makeability faded, there was greater interest in more integrated approaches in environmental policy. The State became convinced that it should approach environmental issues in relation to each other, rather than pursuing a compartmentalized policy of remediation. However, it took some time before this was put into practice.

The 'maturing' of environmental policy went hand in hand with the development of instruments to give shape and structure to it.[9] In the early years of environmental policy, it was usually environmental standards that were embraced as the solution to environmental issues. The realization gradually dawned that ad-hoc approaches to urgent environmental issues would cause environmental policy to become compartmentalized. In the mid-1980s there was a shift of emphasis in environmental policy, and the remediation approach was partly abandoned. The shift was a result of the wish to align the various policy lines. In the Environmental Policy Integration Plan (PIM; *Plan Integratie Milieubeleid*), integration is seen as a condition for effective policy. The principles of this plan proved to be highly

6 Even now, issues still regularly come to light that were seen thirty years ago as the goal of remediation policy (see e.g. Schmit 2005).

7 The belief in a makeable society is largely a concept of socialism and social democracy. In the environmental policy of Den Uyl's Labour (PvdA) cabinet, it was mainly expressed in the desire to control and manage environmental issues by means of technical measures.

8 The policy document on ambient environmental standards (*Nota milieuhygiënische normen*) set out the relevant standards for environmental policy. The document distinguishes between different types of standard, including *quality standards*. It is these standards, which relate to the physical condition of an area, that have consequences for spatial planning.

9 Before environmental policy took on a structural form, the Nuisance Act (*Hinderwet*), which dated back in various forms to before 1875, was the only policy instrument for minimizing environmental nuisance.

decisive for the developments in environmental policy.[10] The shift also placed other demands on the instruments for supporting environmental policy. Hence the need arose in this period for methods that would make it possible to harmonize environmental and spatial considerations in practice. The 'Compact City' spatial-planning policy,[11] in which spatial and environmental objectives increasingly clashed,[12] undoubtedly contributed to this need (Bartelds and De Roo 1995).

The response to this was *area-specific environmental assessment methods*. Perhaps the most widely known of these is the environmental zoning method[13] of the Association of Netherlands Municipalities (VNG). In 1986 the VNG compiled a list of almost every type of business and the recommended distance between the companies in question and a quiet residential area. The aim of the VNG was to provide a practical tool to help policymakers plan for environmentally harmful functions. The method was a success, and the fourth version is now available (VNG 2007).

However, the way in which the original VNG method was structured leaves little room for nuance and consideration. This means that it cannot be used in complex situations involving large-scale enterprise and several sources of environmental burden. The VNG method is a welcome tool for relatively straightforward situations with clear causes and consequences. One of the reasons that Integrated Environmental Zoning (IMZ; *Integrale Milieuzonering*) was introduced in 1989 was to compensate for the shortcomings of the VNG method. The IMZ method was designed to produce an integrated contour of environmental load around an area with large-scale, multiple sources of environmental load. The method classified and standardized various types of environmental load. This made it possible to compare them and, using a cumulation method, 'add up' the loads to obtain an

---

10 The first National Environmental Policy Plan (NEPP-1; TK 1989) built on the principles set out in 1983 in the PIM (VROM 1983). This relates mainly to internal coordination. The stimulus for external integration (i.e. coordination between environment and spatial planning) was largely missing.

11 The concept of the compact city was introduced at the end of the 1970s as a solution to the increasing environmental pressure on rural areas (Bartelds and De Roo 1995). It soon became apparent that the proposed solution – the concentrating and combining of functions in cities – led to increasing pressure on the environment in urban areas.

12 The term environmental/spatial conflict was introduced to describe these situations. Originally, this term referred to conflicts between industry and residential developments (Borst et al. 1995), but it is generally used for 'issues of environmental quality and the spatial planning in an area or location that conflict with each other in some way' (De Roo 2001; 7).

13 Environmental zoning focuses on environmentally sensitive or environmentally harmful functions. In the case of environmentally sensitive functions, zoning is designed to protect quiet residential areas or nature conservation areas, for example. The purpose of zoning is to control the impact of these functions in an acceptable way (VNG 1999). In addition, zoning creates certainty for businesses: they can continue to operate within the specified environmental zone.

integrated value for actual situations. Although the term 'integrated' suggests otherwise, the IMZ method is the classic product of a time when standards-based thinking was seen as the only route to success. Tensions between centrally imposed environmental standards and the possibilities for spatial development at local level are partly to blame for the fact that the method never progressed beyond the 'provisional system' stage.

The supposedly destructive character[14] of the IMZ method and the discussions relating to it undoubtedly contributed to the shift in environmental policy that took place in the mid-1990s. Standards, which were still embraced at the beginning of the 1970s as the solution to environmental issues, are no longer sacred (compare e.g. De Roo 1999). Environmental quality is no longer expressed only in quantitative standards but is more often described qualitatively in terms of liveability. A similar shift can be seen in the methods for environmental externalities developed in the second half of the 1990s. The Rotterdam method 'Milieu op z'n Plek' (A Place for the Environment), for example, is based on a environmental commitment in the form of a locally formulated minimum required quality and a target quality (Municipality of Rotterdam, 1998). Environmental standards are still used, but in a less restrictive and prescriptive way; the standards serve as a guideline. A similar approach is used in the later LOGO[15] and MILO methods.

The shift in environmental policy described above led to a change whereby central frameworks were replaced by greater policy freedom at local level. Local authorities were given more opportunities and greater responsibility with regard to policymaking for the living environment. These developments are categorized under the heading 'decentralization' (Kamphorst 2006, De Roo 2004), making it possible – as Secretary of State Van Geel claimed at the aforementioned 'Kwaliteit van de leefomgeving' conference – to formulate 'an ambitious and attractive environmental policy' (Van Geel 2004).

We have now sketched the development of a policy field that has been in almost constant flux from the start. In that time, priorities have been continually adapted to the circumstances in which the Dutch government found itself. This has produced not only a colourful mosaic of approaches in environmental policy,

14 If planning consequences were to be linked to the IMZ method, large parts of the Drecht cities and Arnhem, for example, would have to be demolished. Yet the method is not always as negative as it is often presented. If it is applied in a different way, i.e. in a less prescriptive and more informative way, it certainly has potential. We will discuss this in Chapters 4 and 6.

15 Like MILO, LOGO (the Dutch acronym for Local Area Typology and Quality of Life) is a typical product of its time. LOGO was developed by the Environmental Services for the Rijnmond region based on experiences with the Rotterdam 'A Place for the Environment' method (DCMR Milieudienst Rijnmond 2004). Although there are clear differences between LOGO and the original MILO method, LOGO was the model for the further development of MILO. This will be discussed in Chapter 8.

but also a diverse collection of methods for environmental externalities. And the developments continue.

1.2 A Changing Policy Field: Decentralization

Environmental policy in the Netherlands is currently undergoing several visible changes. These changes can in fact be categorized under the heading 'decentralization'. Decentralization involves the transfer of responsibility – in this case with regard to the quality of the environment at local level – from central government to lower levels of government. It is a process that is being driven by discussions regarding the central role of the Dutch government, and for which several reasons can be given. De Roo cites the current aversion to regulatory zeal as one of the main motives. In fact, this is such an important motive that 'energetic efforts are being made to break down the existing generic and restrictive policy' (2004; 1). But breaking down the existing policy frameworks will also destroy the certainties that go with them. This is certainly the case if it is not clear where developments are leading, as several authors have pointed out (e.g. De Roo 2004, MNP 2004, Kamphorst 2006, Bouwer 1998).

In his consideration of the future of environmental policy in the Netherlands, De Roo (2004) expresses reservations about the decentralization process. He believes that there is too little control over the process, which means it is not possible to predict the position of environmental policy within the broader policy for the living environment. The Netherlands Environmental Assessment Agency (MNP) has also expressed concerns about current developments. Those concerns relate mainly to the use of new definitions of quality and the lack of related quality objectives that are uniform and measurable (MNP 2004). It also points to the dangers of an 'invisible environmental policy within a broader comprehensive environmental policy' (Kamphorst 2006, Bouwer 1998), whereby it will not be possible to defend environmental interests sufficiently in spatial planning processes. Van Geleuken and Baartmans warn that, in the discussion on centralization, 'little attention is paid to instruments and mechanisms that encourage local authorities to improve the quality of the environment' (2004; 30).

This climate, in which there are no policy certainties and the consequences of policy shifts for future policy are unclear, will undoubtedly affect the development of methods for environmental externalities. At the same time, however, new perspectives are emerging. New rules for local environmental policy can also lead to a situation in which methods for environmental externalities are used in a different way than their designers originally intended. An illustration of this is the transformation that the IMZ method, so maligned in the Netherlands, has undergone in the United States. Primarily a top-down method in the Netherlands,

IMZ (known in the United States as the BAEL Profile)[16] has been adapted so that its main purpose is to provide non-governmental parties with meaningful information about the current environmental situation, without this necessarily having planning consequences. The information, so the theory goes, can contribute to 'communicative action' by organized interest groups and groups of citizens in their quest for improved liveability.[17]

1.3 The Importance of Methods for Environmental Externalities

It is evident that methods for environmental externalities are essential to Dutch environmental policy. The timely integration of the environment in the (spatial) planning process can help to find an effective solution for issues or to improve liveability for local communities. Methods for environmental externalities have proved a valuable instrument for this in recent years. This is borne out by the number of methods for environmental externalities and the practical experiences with these methods. In terms of substance, these methods can help to give environmental interests a useful and therefore fully acknowledged place in the spatial planning process.

It is not only in the past that spatial planners have seen the environment as restrictive and constraining. Research has shown that today, too, the spatial planning sector is somewhat suspicious of environmental issues (Spreeuwers et al. 2007, Bouwman et al. 2005). This can be changed if the right environment-aware method is used at the right time. Environmental interests are no longer a sideline issue; they are part of a shared responsibility. The environment can play a constructive role in the spatial planning process,[18] and high-quality solutions to environmental issues can be developed. We must not lose sight of the fact that a high level of environmental quality can make a real contribution to a positive, healthy quality of life. And on all fronts there are benefits to be had from methods for environmental externalities. Practical experience with a number of methods has shown, for example, that lines of communication between the various actors

16 This method is used to involve citizens in local environmental policy. Elsewhere in the United States too, projects that use such methods are being set up on a bottom-up basis. In Seattle, for example, a method known as Sustainable Seattle (Seattle Planning Department 1994) is in use. Sustainable Seattle is a non-government programme designed to enable the local population to become involved in improving the quality of life in the area. One part of the programme is the City of Seattle Indicators Project, the aim of which is to develop indicators for liveability in the region. A number of these indicators can be found in local-government policy documents (Miller 2004), which shows how influential this method is.

17 This phenomenon will be discussed in Chapter 9, along with 'environmental atlases'.

18 In 2003, the Haaglanden urban region developed a method called MIRUP (the Dutch acronym for Environment in Spatial Planning) in which environmental considerations are central to the planning process. The method provides 'a content-related and process-based foundation for all manner of sustainability aspects' (Stadsgewest Haaglanden, 2003; 5).

can be shortened considerably in comparison to similar situations in which such methods are not used. Actors from different backgrounds can suddenly 'speak the same language' if they all consider the issue using the same method.

Recent developments relating to the City & Environment approach are interesting in this context.[19] In 2006, the interim City & Environment Act came into force, opening up the approach to all municipalities in the Netherlands. For example, local authorities can now relax environmental regulations – under strict conditions. Environmental standards are no longer prescriptive in all situations but are a guideline from which authorities can deviate if they give good reasons. Methods for environmental externalities can be used to substantiate the arguments that should lead to approval or rejection of the relaxation. The MILO method even stipulates the relationship that needs to exist between MILO and City & Environment in order to substantiate deviations from environmental standards (VNG et al. 2004).

In addition to the above examples from policy practice in the Netherlands,[20] there is another development that underlines the importance of methods for environmental externalities. The environmental policy of the European Union is becoming increasingly important in the national policy of the individual Member States (see e.g. Van Ravesteyn & Evers 2004). This relates not only to the influence of the various directives,[21] but also to the European focus on the urban environment. As a follow-up to the Sixth Environment Action Programme of the EU, the 'Thematic Strategy on the Urban Environment' has been developed. The strategy emphasizes the importance of sharing knowledge and best practices EU-wide in the field of urban environmental management. The Dutch 'best practices' (read: methods for environmental externalities) summarized in this book are in line with this strategy.

The above illustrates the value of methods designed to support local or regional environmental policy and of methods that facilitate the alignment of environment and spatial planning. Dutch initiatives in local and regional environmental policy

---

19 The approach has three steps: tackling problems at source, creative solutions within the law, and relaxing the rules. In the period from 1997 to 2004, 25 local authorities in the Netherlands experimented with the City & Environment approach (see also VROM 2003a). An evaluation study has shown that the approach can contribute to a more economical and effective use of the spatial environment (VROM 2004), which means that a more liveable environment is within reach.

20 Support for the City & Environment approach has been discussed above. But there are numerous other ways in which methods for environmental externalities can play a role. Examples include the Strategic Environmental Assessment (SEA) or MER Plan providing for the timely analysis of the environmental impact of plans or programmes (VROM 2004a).

21 Strategic Environmental Assessment, mentioned above, is also the result of a European directive, namely 2001/41/EC, which came into effect on 21 July 2004. The purpose of the directive is to ensure that Member States identify and assess the environmental consequences of certain plans and programmes in advance.

are followed with interest worldwide. In contexts ranging from the City & Environment approach to the aforementioned transformation of the IMZ method into the BAEL Profile, Dutch initiatives have often served as an example.

1.4 A Framework for Methods for Environmental Externalities

A great deal has happened during thirty years of environmental policy and twenty years of methods for environmental externalities. Initially, environmental policy was above all generic and prescriptive, and the first methods for environmental externalities reflected this approach. Environment-protection standards were used in order to establish a minimum quality requirement for the country as a whole. Today, environmental policy has been and is being decentralized, and this is bringing new challenges and opportunities for the future. The decentralization process inevitably involves a change process, which in turn leads to shifts in policy principles, among other things. The interest in different quality definitions is an example of such a shift. In later policy documents and research reports[22] the drive for a high standard of liveability is increasingly elaborated in terms of minimum required quality (limit values) and desired quality (target value).[23] New methods for environmental externalities are being developed within these altered policy frameworks. The LOGO and MILO methods, for example, work with quality profiles comprising a basic required quality and a desired quality. Seen from this perspective, the methods for environmental externalities are in step with the transition that Dutch environmental policy is undergoing. In the recently developed methods, the prescriptive approach has made way for an approach in which all factors are considered,[24] and issue-specific aspects are central.

In short, the decentralization process is leading not only to new principles and approaches in environmental policy, but also to new related methods for environmental externalities. The prescriptive environmental-zoning methods can

22 For example NEPP-4 (VROM 2001), the National Policy Document on Spatial Planning (*Nota Ruinte*, VROM et al. 2004), an evaluation report on City & Environment projects, the report on the Health & Environment Action Programme (VROM 2006a), the Future Environment Agenda (*Toekomstagenda Milieu*, VROM 2006b) and the Strategic Environment Agenda of the Interprovincial Council (IPO 2007).

23 However, various studies by the University of Groningen show that local authorities do not always know how to apply these concepts (see e.g. Bouwman et al. 2005, Spreeuwers et al. 2007). This leads to situations in which 'the limits of the standard are sought' [i.e. they try to see how far they can go] (Bouwman et al. 2005; 136).

24 Every issue has unique aspects, which means that a tailor-made approach may be required to solve it. This can be realized with this approach. Deviation from standards is possible subject to certain conditions. A condition for this method is that the overall 'balance' of the considerations must be positive. This means, for example, that there must not be a negative impact on liveability. The City & Environment approach is an example of this type of policy.

be seen as products of centrally directed environmental policy. The decentralization process has created more room for considering the situation-specific aspects of spatial issues. This view of matters leads to the observation that, as a result of the – more or less – parallel development of Dutch environmental policy and methods for environmental externalities, there now seems to be an environment-aware method for almost every type of issue.

Municipalities in the Netherlands are responsible for implementing environmental policy. It is important that they have sufficient insight into the background, features and mechanisms of the various methods for environmental externalities. Only then, it is assumed here, is it possible to make an informed choice in favour of a particular method. In the past, and the more recent past, several attempts have been made to put this into practice. In 1995, the study 'Afstemming door inzicht' (Coordination through Insight) was carried out (Humblet & De Roo 1995). The study analysed and compared eight environmental assessment methods that were widely known and had great potential. Since then, other studies have been carried out (see e.g. Evertse et al. 2003, VROM 2003b, VROM 2002a). Each of these studies provides insight into the functioning and scope of application of the different methods.

What is lacking, however, is a comprehensive framework for examining the various methods for environmental externalities in relation to each other *and* to environmental policy. This book is an attempt to fill that gap.

The book therefore has more than one goal. In the first place, it provides the reader with insight into sixteen leading methods for environmental externalities that have been developed over the past twenty years. As mentioned, a situation has arisen in which there is a method for almost every type of issue. It can be difficult for policymakers to decide on the most appropriate method for the issue in question. This selection comprises classic environmental-assessment methods, which are still useful in decentralized policy practice, as well as methods for environmental externalities developed more recently. For environmental officers and spatial planners working for municipal and provincial authorities and environmental agencies, this insight can be essential to formulating policy for the living environment.

In the second place, this book argues that there is more than one way to apply methods for environmental externalities. The selection of methods for environmental externalities is analysed by means of a comparison model, against the background of developments in environmental policy. By using the comparison model, we can shed light on how methods for environmental externalities can be used for purposes other than those for which they were originally designed. In this context we again refer to the metamorphosis of the IMZ method from a framework-setting instrument into the BAEL profile, a participative instrument that is largely designed to provide meaningful information on the current (environmental) situation. The latter method is intended to support citizens with information in holding polluters accountable.

1.5 Structure of this Book

This book about methods for environmental externalities is divided into three main sections. The first section contains three chapters. Chapter 2 – building on Sections 1.1 and 1.2 – outlines the development of environmental policy in the Netherlands. It deals with the main features of the policy and the social context in which it evolved. An understanding of the historical development of environmental policy is important in the light of the close relationship between environmental policy and the methods for environmental externalities that have evolved from it. The developments will be dealt with chronologically and from time to time we will 'digress' to examine the methods for environmental externalities that have been developed over the years.

The outline of the development of Dutch environmental policy will consider the various shifts that can be identified. Each shift has led to new principles for environmental policy, and consequently to new methods for environmental externalities. In Chapter 3, based on the characteristics of these policy shifts, a model will be constructed for analyzing the methods for environmental externalities. In principle, the model is designed to do three things. In the first place, it is designed to assess the methods individually. In the second place, the methods can be compared using the model. Finally, the model's simple and direct structure means that it can be used to compare other methods for environmental externalities to those presented in this book.

The selection of methods for environmental externalities is introduced in Chapter 4. In order to retain the overall picture, the methods are divided into categories according to their common characteristics, principles and mechanisms. The categories are as follows: zoning methods, checklist methods, quality profiles, 'merge' methods and information methods. There are three methodologies that do not readily fit into any of the categories. These will be put into an 'Other' category.

In Part B, (Chapters 5 to 10) the comparison model is applied to the selection of sixteen methods for environmental externalities. The selection consists of classic environmental-assessment methods as well as methods developed more recently. The selection therefore reflects the various emphases of environmental policy in the Netherlands. Each chapter deals with a separate category.

The environmental zoning methods are explained in Chapter 5. The methods in this category are the zoning list (*Handreiking Bedrijven en milieuzonering*) of the Association of Netherlands Municipalities (VNG) and the Provisional System for Integral Environmental Zoning (VS-IMZ). In the category Checklist Methods (Chapter 6), the Amsterdam Environmental Performance System (*Milieuprestatiesysteem Amsterdam*), the *Milieuplaberum* (planning & decision-making process for spatial measures) and Health Impact Screening will be discussed. Chapter 7 will focus on quality profiles. This category comprises the Utrecht 'Bandwidth' method (BBM, *Bestuurlijke Bandbreedte Methode*) and the neighbourhood-oriented environmental targets (*Wijkgerichte Milieustreefbeelden*) used in the city of Groningen. Chapter 8 discusses the 'merge' methods.[25] These are the LOGO, MILO and MIRUP methods. Chapter 9 discusses the 'information' methods.[26] The two methods in this category are the Deventer Environmental Atlas (*Milieuatlas Deventer*) and the Groningen Environmental Atlas (*Milieuatlas Groningen*). Chapter 10 discusses three methodologies that, in terms of principles and mechanisms, do not fit into any of the other categories. These are the Environment Maximization Method (*Milieu Maximalisatie Methode*), the ROMBO tactic and the *Stolpmethode*.

Part C has two concluding chapters. In Chapter 11, the selected methods are compared using the model described in Chapter 3. Chapter 12 consists of concluding remarks.

---

25 The term 'merge methods' is used in this book to denote methods that bring together and integrate different policy tracks (spatial planning and the environment). Other terms can also be used. SenterNovem uses the term 'coat-rack methods' (*kapstokmethoden*) because of the different sub-instruments that 'hang' on them (SenterNovem 2007).

26 The order in which the methods are discussed is explained in Chapter 4.

PART A
Context and Conditions

Chapter 2

Developments in Environmental Policy: Standards, Remediation Politics and Integrated Environmental Management

Environmental problems are timeless. This is illustrated by the Roman philosopher and statesman Lucius Annaeus Seneca. In a letter he wrote on the conditions in ancient Rome: 'No sooner had I left behind the oppressive atmosphere of the city and that reek of smoking cookers which pour out, along with a cloud of ash, all the poisonous fumes they've accumulated in their interiors whenever they're started up, than I noticed the change in my condition at once.'[1] Seneca's observations apply not only to Ancient Rome: 'For almost as long as there have been cities they have been polluted' (Brimblecombe and Nicholas 1995; 285).

A similar claim could be made for the Netherlands too. However, until well into the 1960s politicians paid hardly any attention to environmental problems. There was a certain amount of legislation and regulation, but no structured environmental policy. This began to change in the course of the 1960s. Rachel Carson's book *Silent Spring* (1962) had already been published by then. In 1968, an article entitled 'Tragedy of the Commons' by Garrett Hardin appeared in the journal *Science*. The article was about the mechanisms that eventually lead to environmental problems. However, it was not only publications that raised public awareness. Environmental disasters also played a role. This process appeared to reach its zenith around 1970. A political response was not long in coming.

That response took the form of the Priority Policy Document on Pollution Control (*Urgentienota Milieuhygiëne*), which was published in 1972. Dutch environmental policy had begun to take shape in a structural sense. In some respects, the Priority Policy Document marks the birth of Dutch environmental policy. Developments then followed in rapid succession. In retrospect, we can divide these into a number of periods, each of which produced its own procedures, instruments and methods. The policy principles of the various periods are also reflected in methods for Integrating Environmental Externalities in Urban Planning and Design (hereafter referred to as 'Methods for Environmental Externalities').

The first of these methods was not introduced until the mid-1980s, however. Environmental policy had then existed for ten to fifteen years, and had evolved into a policy sector with its own procedures and standards. The first Methods for

1 Extract from letter 104 written by Seneca as translated by Costa in 'Seneca: 17 letters' (1988).

Environmental Externalities had the same characteristics as early environmental policy; above all, they set standards and frameworks. As the principles of environmental policy changed, similar changes occurred in Methods for Environmental Externalities. In later methods, therefore, the focus also shifted to support and process optimization. During the period in which this change took place, many wide-ranging Methods for Environmental Externalities were introduced. In order to understand the basic principles of these methods, it is important to understand the period in which they were developed.

This chapter focuses on developments and changes in Dutch environmental policy. These will be discussed chronologically, with reference to the methods of the period in question. In Section 2.1 we will discuss changes in policy and policy views. Particular emphasis will be placed on developments leading to a new or modified course in environmental policy. The periods of environmental policy are discussed in detail in the following four sections. A conclusion follows in Section 2.6.

2.1 Periods in Dutch Environmental Policy

Changes in policy and policy implementation do not usually take place overnight. It is usually a long and troublesome process[2] preceded by growing dissatisfaction with the prevailing policy views. Sometimes there are clear reasons for making policy changes. In the case of Dutch environmental policy, these reasons date from the beginning of the 1970s. At that time there was a strong increase in environmental awareness on an international level. Environmental issues were increasingly discussed in the media. Large-scale disasters involving oil tankers and chemical plants contributed to this, but well-known publications such as *Silent Spring* and, later, the report of the Club of Rome also resulted in the environment becoming an issue on the social agenda.

This societal interest also made it clear to the government that intervention was desirable, if not essential. Eventually, the environment was given priority on the *political agenda*. This point is often seen as the birth of environmental policy in the Netherlands. Developments in environmental policy then followed in rapid succession.

Depending on the perspective from which it is considered, the history of Dutch environmental policy can be divided into a number of periods. This book, *Smart Methods for Environmental Externalities*, focuses on the influence that a period in environmental policy has (or has had) on the development of methods and instruments. The result is a division in four periods. Each period is characterized by a particular approach to the environment and how it is regarded in relation to other policy fields. Figure 2.1 is a schematic representation of the four periods,

2 In their book *Policy Change and Learning* (1993), the American policy scientists Sabatier and Jenkins-Smith show it can sometimes take decades for policy changes – and intended changes – to be fully implemented.

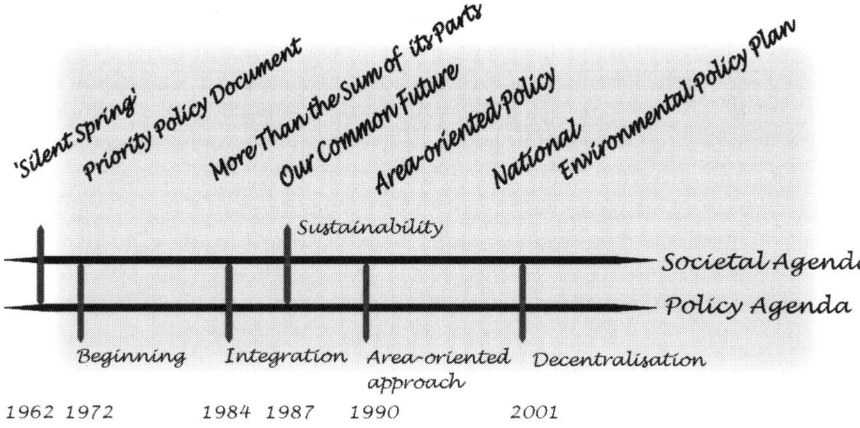

Figure 2.1 Periods in Dutch environmental policy

indicating in each case the influential publication that heralded the beginning of a 'new period'.

In the first period, which began in about 1970, environmental policy was developed from scratch. The first policy documents and approaches clearly show that the government was confronted with new material; it was looking for the 'right' approach. Partly for this reason, the very first environmental policy was governed by a short-term vision, a compartmentalized approach and with a focus on reactive policies and clean-up operations. In this first period, policy was shaped mainly by the Priority Policy Document (VM 1972) that was published in 1972.

In 1984, the policy document 'More than the Sum of its Parts' (VROM 1984a) was published, heralding the beginning of a second period. The publication of this policy document was preceded by a period in which it became clear that policy must be explicitly coordinated and cohesive if problems are to be tackled effectively. Internal integration is a key concept in that respect. As a result of the continued development of environmental policy, policymakers as well as policy implementers were confronted with a complex system of legislation, regulations and procedures. This time period also marks the first development of Methods for Environmental Externalities which were developed in an attempt to retain an overview.

In the third period, the focus shifted from internal integration to external integration. The beginning of the third period is marked by the publication of the Action Plan for Area-oriented Environmental Policy (*Actieplan Gebiedsgericht Milieubeleid*, TK 1990). At the end of the 1980s it became clear that combining and coordinating various elements of environmental policy as was done with internal integration was not enough to ensure that it would be effective. The focus was therefore shifted from internal integration to external integration which was focused on coordinating and integrating environmental policies with other policies such as economic, social and spatial policies. Particular emphasis was placed on coordination with spatial planning during this period.

The year 2001 saw the publication of the fourth National Environmental Policy Plan, or NMP4 (VROM 2001), which can be seen as the beginning of a fourth period in Dutch environmental policy. In NMP4, environmental policy is based on the concepts of liveability and what is called in the NMP4 'the quality of the living environment'. It also marks a period of decentralization and deregulation in Dutch environmental policy so as to advance the role of local authorities in the pursuit of liveability and 'the quality of the living environment'. The Methods for Environmental Externalities developed in this period are also based on these concepts.

In the following sections, the emphasis is on two aspects. First, environmental policy itself. The discussion is structured according to the periods identified. The second aspect is the Methods for Environmental Externalities that were developed during these periods. The first methods were introduced in the mid-1980s and were typically conditions setting in character. The conditions are being represented by quantitative standards, issued top down by the national government. New methods appeared in subsequent years, and the emphasis shifted each time. The method and principles of these methods derive directly from prevailing views in environment policy, as we shall see.[3]

2.2 A Tangible Beginning: The Priority Policy Document on Pollution Control

Before 1970 there was no cohesive environmental policy. If environmental issues were tackled at all, this was done in a fragmented and unstructured way. Moreover, environmental issues were not dealt with until problems arose; there was no such thing as preventive policy. Eventually the government came to realize that this approach to the environment was not very effective or future-oriented.

Yet it took a long time for the public sector to reach this conclusion. This process of increasing public awareness appeared to reach its zenith around 1970. Environmental awareness had also increased because people were confronted with environmental pollution in their own surroundings. The establishment of the Directorate-General for Environmental Protection (DGMH) in 1971 within the then Department of Public Health and Environmental Protection was the first institutional and organizational foundation for Dutch environmental policy. The tangible beginning came a year later with the publication of the Priority Policy Document, which contained the first concrete proposals for tackling environmental problems in a structured way. The chosen approach, whereby environmental

3 The policy lines and principles discussed in this chapter form the foundation of a comparative framework for Methods for Environmental Externalities. The framework is explained in Chapter 3.

policy was linked with public health, determined environmental policy to a large extent in the 1970s.

The way in which policy on the environment and environmental protection was formulated in the Priority Policy Document was above all sectoral and framework-setting. As a result, separate legislation and regulations were drawn up for the various environmental compartments (water, air and soil), supported by a framework of standards. In 1976, this was followed up with the Policy Document on Ambient Environmental Standards (*Nota milieuhygiënische normen*, VM 1976), which later became the basis for quantitative and evaluative Methods for Environmental Externalities. But it would be another ten years before these were developed. The fragmentation of policy into compartments and sectors resulted in a wide range of licensing systems, procedures and implementation mechanisms. Consequently, it proved necessary to split environmental issues into a number of relevant smaller questions that fit within one of the environmental compartments. Then, responsibilities are spread among various policy sectors. A result of this approach was that issues were not solved but were rather passed on to other departments and agencies or simply deflected from one compartment to another.

The publication of the Priority Policy Document was followed by a proliferation of various types of sectoral legislation.[4] This legislation focused on specific environmental threats (waste, radiation, noise). One of these laws led to the introduction of *zoning*, an instrument based on environmental protection standards and physically separating environmentally intrusive and sensitive functions. Although emerging within noise policies, it was not introduced in the Noise Abatement Act (*Wet Geluidhinder*) but in the amended Aviation Act (*Luchtvaartwet*) of 1978 (for a discussion, see De Roo 2001). With the concept of environmental zoning, a direct bridge between environmental standards and spatial planning was for the first time created. It would however take a while before environmental-zoning methods were introduced (see Box 2.1).

Towards the end of the 1970s, there was an increasing awareness of the disadvantages of the sectoral approach to environmental issues. In practice, it became evident that compliance with a particular piece of sectoral legislation could have a negative effect on other policy sectors. It became clear that the implementation of sectoral policy did not solve environmental issues in every case, but resulted in the deflection of pollution to other compartments, areas or to next generations. Partly as a result of this, there was an increasing need for integrated policy and fewer government regulations. That need also encompassed a call for instruments to support the implementation of such a policy.

---

4 The Priority Policy Document was preceded by two pieces of sectoral legislation: the Pollution of Surface Waters Act (WVO, *Wet verontreiniging oppervlaktewateren*) in 1970 (Smit 1989) and the Air Pollution Act (*Wet Luvo*, or: *Wet inzake de luchtverontreiniging*) in 1970 (Michiels 1989). The Priority Policy Document announced legislation for noise pollution and soil pollution (VM 1972).

Box 2.1 Spatial layout in the Netherlands

The way in which space is utilized is very important in the densely populated Netherlands. In an absolute sense, the amount of space available does not change, but the demand for space increases all the time. In order to ensure that space is properly allocated, planning is required. This happens at all levels of the public sector. At the local level, the most important instrument in this regard is the zoning plan, or land-use plan (for an explanation see Voogd 1999 and Spit & Zoete 2002). The zoning plan is a unique type of spatial plan. It is the only type of spatial plan that is legally binding for all parties. For a long time, however, zoning plans did not contain environmental requirements.

The Spatial Planning Act (WRO, Wet op de ruimtelijke ordening) stipulates how spatial plans, and therefore zoning plans, are drawn up. The current act dates from 1965 and has been amended many times. A bill to amend the WRO was submitted to the Second Chamber of the Dutch Parliament in 2003, which has come into force in 2008. Among other things, the new WRO gives local authorities greater scope for including environmental quality requirements in zoning plans* (see also VROM 2003c, VROM 2006b).

* In the 1990s, the municipal authority of Amsterdam attempted to include explicit environmental requirements in zoning plans (see e.g. Timár 2005). However, the Council of State forbade this and declared it to be in contravention of the WRO. This is discussed in Chapter 6.

Yet it was some time before the first Methods for Environmental Externalities were introduced in environmental policy. In this first period, as various policy documents show, the emphasis was on structuring and establishing environmental policy. The shortcomings did not become apparent until the policy was in place. Only then did the government start to consider coordinating and integrating the various lines in environmental policy in order to create a consistent, balanced and effective whole, thereby reducing the procedural and organizational fragmentation of environmental policy. This aim for internal integration heralded a second period in Dutch environmental policy.

2.3 On the Way to Internal Integration

In 1980, the provisional Committee for Long-Term Developments in Environmental Policy was set up[5] with the aim of 'designing an organizational framework, in order to provide the Minister for Public Health and Environmental

5 The provisional Committee for Long-Term Developments in Environmental Policy (also known as the Clarenburg Committee or CLAT) witnessed the inception of the permanent Committee for Long-Term Environmental Policy (CLTM) set up in 1988

Protection as quickly as possible with cohesive information about the most recent insights into developments in the medium and long term that may be important for environmental policy' (CLTM 1990; 3). The reference to providing cohesive information is the most striking aspect of this description. This point of departure shaped the further development of environmental policy and influenced the formulation of the first environmental-assessment methods.

In the meantime, the Second Chamber of Parliament had taken the position – following the advice of the De Wolff Committee (1970, see also VWRR 1975) – that environmental policy should have a more framework-setting character (facet policy) rather than a sectoral character (see also Voogd 1999). As a result of this shift, environmental policy would be more influential for the development of policy sectors. It was assumed that this would make it possible to achieve an integrated approach in environmental policy (also Box 2.2). The integration process that followed proved to be difficult and did not lead to the consistent policy that the government had initially aimed for.

Box 2.2 Sector and facet planning

The distinction between sectoral and facet planning was formulated in 1970 by the De Wolff Committee in order to clarify the relationship between different policy fields. Above all, therefore, it is an analytical distinction (Spit and Zoete 2002) or conceptual model (Hidding 1997).

The De Wolff Committee described sector planning as 'the concrete programming of a branch of government activity (corresponding to a department or agency of a department) geared to ensuring that the activity is carried out as efficiently as possible. This planning is characterized by its more or less "technical" character.' (1970; 6). In contrast, the aim of facet planning was 'to integrate all government activities from a single perspective' (De Wolff Committee 1970; 6). In line with the recommendations of the De Wolff Committee, the distinction between sector planning and facet planning was translated into organizational terms for the government apparatus (see also Hidding 1970). The various planning offices, such as the Netherlands Bureau for Economic Policy Analysis (CPB, Centraal Planbureau) and the Netherlands Institute for Spatial Research (RPB, Ruimtelijk Planbureau) formerly the National Spatial Planning Agency – can thus be regarded as 'facet planning agencies'.

In 1983 the Environmental Policy Integration Plan (PIM, *Plan voor de integratie van het milieubeleid*) was drawn up to coordinate environmental policy and spatial planning. The PIM proved to be an interesting policy document for several reasons. In the first place, the PIM moved away from the compartmentalized approach to environmental problems that was advocated in the Priority Policy Document.

by the Central Council for Environmental Protection (CRMH, *Centrale Raad voor de Milieuhygiëne*).

Whereas the Priority Policy Document approached the problems relating to air, water and soil separately, these aspects were no longer regarded as separate in the PIM. Concrete guidelines for realizing this integration followed a year later with the publication of the government policy document 'More Than the Sum of its Parts' (VROM 1984a).

The second reason why the PIM is interesting in this context relates to Methods for Environmental Externalities. The PIM was the first serious consideration of the possibility of setting differentiated standards and working with a 'bandwidth' with an upper and lower limit values expressing different ambitions. These principles formed the foundation for the Methods for Environmental Externalities of the 1990s.

In the period that followed the publication of the PIM, in the mid-1980s, the first Methods for Environmental Externalities were introduced. At that time they were known as environmental-assessment methods (Humblet and De Roo 1995). The discussion on the development of an integrated planning system was now well under way. The Association of Netherlands Municipalities (VNG) responded to this and developed a method that created a bridge between environmental standards and spatial planning. The method is an environmental-zoning method that can be used 'systematically to determine the required distances between businesses and environmentally sensitive locations' (VNG 1986; 17). The VNG method was the first integrated environmental-zoning method to be introduced in the Netherlands (for a discussion see Section 5.1) The method consists of a list of company categories and the related average distances between those companies (i.e. environmentally harmful activities) and environmentally sensitive locations such as residential areas or nature areas. The method is entirely in line with the prevailing policy views of the time: framework-setting and internally integrated.

In the mid-1980s, as the government attempted to realize the planned internal integration, a new sense of urgency arose in society. Again, this was due to large-scale environmental disasters (see also Ale 2003) and resulted in the publication of documents describing the worrying state of the environment. Important factors included the disaster involving pesticides at a chemical factory in the city of Bhopal in India, the nuclear reactor accident at Chernobyl (1986) and the Exxon Valdez oil spill on the coast of Alaska (1989). The publication of the Brundtland Report 'Our Common Future' (WCED 1987) also played a key role. The report warned that the environment could not sustain the endless pressure put upon it by mankind, and emphasized the need for sustainable development. The term 'sustainability' would prove to play a major role in the environmental policy of the 1990s.

The renewed sense of urgency was one of the factors that led to the amendment of several fundamental points of environmental policy. A first attempt at change was the publication of the report 'Concern for Tomorrow' (*Zorgen voor Morgen*) by the National Institute for Public Health and the Environment (RIVM). The report was the Dutch response to the Brundtland Report. It was drawn up in preparation for the first National Environmental Policy Plan (NMP1), which mapped out the course of environmental policy in the 1990s. Principles from earlier environmental policy of the 1970s, such as the 'stand still' principle, 'the polluter pays' and tackling problems at source were also the principles of NMP1. The plan also contains new points of departure. The concept of sustainability was also central, in line with the Brundtland Report as well as the RIVM report. In NMP1, the main objective is defined as follows: 'to preserve the impact-bearing capacity of the environment in the interest of sustainable development' (TK 1989; 92).

NMP1 stated that it was not possible to tackle all pollution effectively by using at-source measures. Effect-oriented methods were also considered necessary in addition to at-source measures. However, NMP1 still devoted little attention to the effect-oriented approach. There are a number of reasons for this (see De Roo 2001). From the point of view of environmental protection, at-source policy measures are more effective than effect-oriented policy. Moreover, effect-oriented measures were seen as a 'dying species'.[6] Despite all of this, the government assumed that '(…) effect-oriented measures will still be necessary in the medium term' (TK 1989; 86). This is expressed in NMP1 in the wish to 'promote integrated environmental zoning', an effect-oriented measure *par excellence*. The Integrated Environmental Zoning method (IMZ, *Integrale Milieuzonering*) is based on the idea of creating buffer zones between environmentally intrusive and sensitive functions, so as to ensure that these sensitive functions are not subject to intolerable environmental stress. In this, various environmental stressors are cumulated and expressed as an integrated score underlining the integrated character of the approach. The approach is essentially 'a means of reducing effects and risks at the local level. In this respect it relates to the designated-areas policy' (TK 1989; 52). At the beginning of the 1990s, in order to implement IMZ policy, a method was developed for delineating and implementing integrated environmental zones around industrial activities. That method was the Provisional System for Integrated Environmental Zoning (VS-IMZ) (VROM 1990, for a discussion see Section 5.2).

Unlike the VNG tool, VS-IMZ was supposed to become part of Dutch environmental law; i.e. in section 6 of the Environmental Protection Act (WM, *Wet Milieubeheer*). It never got that far and if the success of VS-IMZ is measured by the number of integrated environmental zones actually defined *and* implemented, then the result is extremely disappointing. The method was used in eleven pilot projects (for a discussion see Borst et al. 1995). An integrated environmental zone was defined and implemented in only one of these projects. Despite this result, VS-IMZ has a number of very interesting features (the method is discussed in Section 5.2, see also Borst et al. 1995, Borst 1996). The innovative aspect of IMZ policy was the way in which it dealt with several categories of environmental pollution together and formulated an integrated environmental quality.

---

6 There are various reasons for this. The environmental policy of the 1970s can be considered as pre-eminently effect-oriented and is not regarded as successful. Another cause of the 'extinction' of effect-oriented policy is the fact that, from the point of view of environmental protection, at-source policy is considered to be more successful than effect-oriented policy (TK 1989).

The first National Environmental Policy plan did indeed propose integrated environmental zoning as a bridge that would link environmental policy and spatial planning, but further encouragement to integrate with other fields of policy was largely lacking. The next section focuses on the third period in environmental policy in the Netherlands. In this period, integration with related policy fields *was* an objective. The period was mainly characterized by external integration and the development of an area-specific approach in environmental policy. This is reflected in the Methods for Environmental Externalities developed at the time.

2.4 External Integration and an Area-Specific Approach

The publication of the Action Plan for Area-Specific Environmental Policy (TK 1990) was the first step towards external integration. It reflected an active approach to better connect spatial planning and environmental policy. The principle of sustainability was clearly reflected in this action plan, too: 'area-specific environmental policy is geared towards creating the environmental conditions that are essential for the sustainable development of the living environment' (TK 1990; 5). The point of departure for the policy was the 'quality of the living environment' (TK 1990; 9), which is considered the concept to connect spatial policy and environmental policy. An important aspect of area-specific environmental policy was to better combine various land uses and activities coexisting within a specific area. During this first phase of area-specific approaches, the 'traditional' instrument of environmental zoning played an important role.

Both the ROM (Spatial Planning and Environment) policy and Integrated Environmental Zoning were pilot projects referred to in NMP1 and the Action Plan for Area-Specific Environmental Policy as possible methods for realizing area-specific environmental policy. However, there are fundamental differences between the approaches taken within these two projects. Integrated environmental zoning proved to be a product that followed the path of the 1980's, focussing on the development of a condition setting policy framework and the internal integration within environmental policy. Hence, the term 'integrated' remained limited to the internal coordination of various aspects of the environment. By contrast, the ROM approach *did* address external integration. Described in the Fourth National Environmental Policy Plan (NMP4) as an integrated area-specific approach at regional level, it did focus on combing environmental aspects with other relevant areas specific qualities such as social, economic and infrastructural qualities. It thus came closer to what the Action Plan for Area-Specific Environmental Policy described as an integrated area-specific approach: 'an area-specific policy whereby the desired quality of the living environment is achieved through a combination of spatial and environmental measures' (TK 1990; 10).

These notable differences between the two approaches in environmental policy evolved more or less simultaneously. They would eventually define the boundaries of what can and cannot reasonably be achieved with environmental zoning. After all, the use of the VS-IMZ method in various pilot projects made it painfully clear what a too-rigid application of generic environmental standards could achieve. Consequently, this method was no longer mentioned in spatial plans from the mid 1990's onwards, reinforced by the fact that the then National Spatial Planning Agency (RPD) regarded the IMZ method as too much of a threat to future spatial development. Municipal authorities confirmed this attitude in expressing their fear of the consequences of the VS-IMZ method for local spatial ambitions.

Following the 'rejection' of integrated environmental zoning, the calls for a less quantitative approach became ever louder. Policymakers began to focus on the possibilities for optimization. One method that appeared to fit the bill was the Bubble Method (*stolp-methode*), which was developed in the mid-1990s. This method offered scope for weighing up and compensating the different forms of environmental stress by means of area-specific standardization. The success of this method was however limited to breaking through dogmatic thinking about environmental standards. It proved to be virtually unworkable.[7] When the method was operationalized, it turned out to be very similar to a zoning method, and that is precisely what the approach was designed to avoid. By contrast, the ROM designated-areas policy gained much more support. This was not just caused by its explicit focus on external integration, but also by its simplicity and the emphasis on local support and tailor-made solutions.

The ROM designated-areas policy was set up as an experiment. Initially, six areas were designated in which there were serious local environmental problems. A number of 'clean' areas were added later, influenced by growing environmental awareness (TK 1988b). ROM areas are characterized by the fact that the problems are highly specific and embedded in a local or regional context. There are also many stakeholders with diverging interests. Policy for these areas was therefore to be based on these features. The ROM designated-areas policy is a decentralized instrument for environmental policy that strongly emphasizes an approach tailored to local circumstances and cooperation between those directly involved. The principles of the ROM designated-areas policy are also found in the Methods for Environmental Externalities developed later in the 1990s: communicative, tailor-made solutions, decentralized, flexible.

These Methods for Environmental Externalities never constituted a complete departure from standards setting environmental policy. The mid-1990s saw the publication of the first in-depth study, involving eight Methods for Environmental Externalities that were known at the time and regarded as promising (Humblet and De Roo 1995). The study revealed the conditions- and framework-setting nature of the methods, which were mainly developed at a time when external integration was a priority. In the study, the methods were therefore referred to as environmental-assessment methods. The methods in question included those already mentioned:

7 In reviewing the method, alignment was sought with prosperity theory, *inter alia*. Partly as a result of this, the gulf between the method and actual practice increased further. The Bubble Method is discussed in Chapter 10.

the VNG method, VS-IMZ and the Amsterdam 'Bubble Method'. Other less well-known methods were also discussed in the study: the Zwolle method, the Groningen environmental-assessment method and the Lelystad Zoning Method.

The decentralization of environmental policy introduced in the ROM designated-areas policy was taken further in the Second National Environmental Policy Plan (NMP2). The main theme of this plan was to reinforce the realization of policy intentions from NMP1 (TK 1993). The aim of NMP2 was to delegate more responsibilities for the environment to lower levels of government. To that end, NMP2 provided a framework for self-regulation at the local level. Consequently, policy was developed at the local level, close to the local stakeholders and circumstances.

A shift in the direction in which environmental policy was developing is evident in the extent to which integrated environmental zoning was rejected and the enthusiasm with which ROM projects were initiated. It was a shift away from centralized top-down governance towards policymaking at the local or regional level. The focus shifted from the generic to the specific. A similar shift is evident in the Methods for Environmental Externalities that were developed from the mid-1990s. The 'hard' assessment of an environmental situation based on framework-setting and generic environmental standards made way for an approach based on area-specific environmental qualities. In 1998, the method 'A Place for the Environment' (MOZP, '*Milieu op z'n Plek*') was developed in Rotterdam (Gemeente Rotterdam 1998). This method is a departure from the framework-setting nature of environmental standards. In this method, legal standards are considered as indicative guidelines. The later LOGO method is based on MOZP.

The period following the publication of NMP2 is therefore characterized by increasing emphasis on the fit of environmental policies and regulations within a context of local conditions, interests and hence, other policy priorities. It marks a shift in environmental policy, from a hierarchical and framework-setting approach to a local and tailored approach and can be regarded as a new governance philosophy in environmental policy (see also VROM 1995; Appendix 1).

The development of this new 'governance philosophy' involves three aspects. First, caring for the environment is a matter not just for the government but for everyone concerned. Second, policy must be implemented based on public and stakeholder participation way. Finally, it must be possible to deviate from set standards – within a certain 'bandwidth'. These principles clearly reflect the wish to make environmental policy part of an integrated comprehensive policy for the physical environment as a whole.

In the years that followed, priority was given to formulating and developing this aim for the various levels of government. A series of policy documents and plans were published that brought an integrated policy for the living environment closer. At central-government level, one such publication was the 'Environment and Economy' policy document (*Nota Milieu en Economie*) (VROM et al. 1997). The document describes the basic steps for integrating the environment and the economy. At provincial level, integrated plans for spatial planning, water management and the environment (POPs) were drawn up and various area-specific projects were implemented (see e.g. De Roo and Schwartz 2001, Oosterhoff et al. 2001). Interesting initiatives at municipal level were the City & Environment (*Stad&Milieu*) projects. This summary is far from exhaustive, but clearly shows the commitment to the theme of integrated environmental policy in all layers of the public sector.

The third period in Dutch environmental policy was thus primarily one in which a broadening of policy occurred. The process of external integration was taken further, and environmental policy appears to have become part of an integrated and comprehensive policy for the overall local environment. The concept of sustainable development played an important role in this. The early 1990s saw the development of the 'Ecopolis' strategy for ecologically sound urban development (Tjallingii 1994, Tjallingii 1995). This strategy, and the later 'Two Networks Strategy' that is based on Ecopolis (Timmermans et al. 2002), were widely applied in later environmental spatial measures. The MZOP method mentioned above and the BBM Method (*Bestuurlijke Bandbreedte Methodiek*) of the Utrecht municipal authority (BRU-projectgroup 1996) were also built on the principle of sustainability. In Belgium, various parties participated in a joint project to develop a 'Sustainability Barometer', which is now used by many local authorities. Sustainability is also embraced at other administrative levels. The European Commission developed a 'Local Sustainability Profile' (EC 2003) to help local authorities to create a sustainable society.

The third period in Dutch environmental policy proved to be a fruitful one in terms of Methods for Environmental Externalities. After a hesitant start in the second half of the 1980s, the development of methods gained momentum at the beginning of the 1990s. With the exception of the occasional 'exotic experiment', such as the Amsterdam 'Bubble Method', it can be said that Methods for Environmental Externalities had found their way into local environmental policy. Towards the end of the 1990s, environmental policy however changed course again. This time, the focus shifted towards the quality of the living environment even more.

2.5 Focus on the Quality of the Living Environment

The Fourth National Environmental Policy Plan (NMP4) marks the beginning of the fourth period in Dutch environmental policy. This policy document begins with an analysis of seven important environmental problems that had not been resolved or dealt with effectively, despite attempts made through previous NMPs (VROM 2001). In the case of local problems, NMP4 aims for policy innovation with the objective to optimize the quality of the living environment. This sounds obvious, but it is not. It implies that attention should no longer be solely focussed on the environment proper (i.e. environmental health and hygiene), but that the environment must be regarded as part of an overall level of quality. It is therefore a statement of principle with far-reaching consequences. The policy proposed in NMP4 does not stand apart (VROM 2001), but aligns with other administrative agreements such as regional covenants, the Investment Budget for Urban Renewal (ISV, *Investeringsbudget Stedelijke Vernieuwing*) and the Major Cities Policy (GSB, *Grote Stedenbeleid*).

The choices made in the NMP4 were in line with the overall development in Dutch polices at the time, in which lower levels of government were granted more powers to formulate policy for the physical environment as they saw fit. In practice, this meant moving closer to a situation-specific and area-specific approach, whereby different levels of quality can be aimed for. In this decentralized approach, issues with a clear local profile are no longer dominated by national governmental policies but, rather, are addressed at a local or regional level.

The developments in environmental policy are described as a shift towards an integrated and area-specific policy. Various experiments from the 1990s onwards illustrate this shift. One example is the City & Environment project (1997–2003) and the modernization of noise-abatement and soil policy. In the City & Environment project, local authorities were allowed to deviate – under certain conditions – from existing environmental standards. In recent years, several Methods for Environmental Externalities have been developed that can fulfil a supporting role in City & Environment procedures. The LOGO method (Section 8.1) and Health Effect Screening[8] (Section 6.3) are examples of these.

In 2006, the Future Agenda on the Environment was published (VROM 2006b). This policy document confirms the choices made in the NMP4 and expresses the current perspective of the national government on current and future environmental policies.[9] One of the priorities of the Future Agenda is to achieve greater coherence between spatial-planning policy and environmental policy. Both forms of policy are geared towards the quality of the living environment. The Future Agenda builds on the NMP4 and re-emphasizes the need for specific instruments designed to achieve this. According to the Ministry of Housing, Spatial Planning & the Environment (VROM), 'formulating environmental ambitions in relation to spatial development also offers possibilities for realizing an optimal quality of life' (2006b; 36). Notably, the document makes no mention of the various methods (LOGO, MILO, MIRUP) that were already in existence *and* have proven to be useful in terms of their contribution to formulating and implementing environmental ambitions (see also Chapter 8). This is especially

8 The City & Environment procedure comprises three steps. In the first two steps, solutions are sought within existing legislation. If this does not produce a satisfactory result, the third step (also known as an 'escape clause') is initiated. In the third step, local authorities must carry out Health Effect Screening (see also Chapter 6).

9 The Future Agenda was not a new environmental policy plan. After the 1990s, the Dutch economy came to a virtual standstill, which meant that plans and measures in NMP4 were postponed. These plans were realized later in the Future Agenda.

surprising as the MILO method was introduced by the Ministry itself in the NMP4 as *the* instrument for decentralized environmental policy.

The aforementioned Methods for Environmental Externalities address the needs formulated in the NMP4 and the later Future Agenda. Yet in the environmental policy of the future there are also opportunities for earlier Methods for Environmental Externalities. In the 1990s, the Amsterdam municipal authority developed the Environmental Performance System (*Milieuprestatiesysteem*) (Section 6.1). This method made it possible to include environmental measures in land-use plans, something that did not please the Council of State at the time. The publication of the Future Agenda seems to create new opportunities for this method. After all, it comes close to a proposal made by the Dutch Cabinet, which wants to give municipal authorities additional options for realizing an optimum quality of life by 'including environmental quality requirements in zoning plans' (VROM 2006; 36).

The possibilities in Methods for Environmental Externalities in future environmental policy coincide to a certain extent with developments in spatial planning. With the Policy Document on Land Use (*Nota Ruimte*), for example, the central government has embarked on a course which shifts focus away from 'setting constraints' (VROM et al. 2004; 36).[10] Instead, the document addresses the growing need for local and regional tailor-made solutions and more interactive planning by the actors involved. The tailor-made approach and interactive planning in particular are also reflected in the most recent Methods for Environmental Externalities. Methods for Environmental Externalities can thus provide a welcome supplement to future environmental policy and spatial planning policy in order to partly or completely fill the gap that was created when the traditional standard-based system was abandoned.

2.6 Conclusion

Looking back at the developments in environmental policy in the Netherlands, one can conclude that environmental policy is a dynamic policy field. A great deal happened between the point at which society became more environmentally aware and the point at which methods were presented that local authorities could use to create a high quality living environment. The 1970s saw the rise of the framework-setting system of standards that, barely thirty years later, were no longer seen as the only route to success. There were developments relating to the integration of policy sectors, a growing emphasis on the specific nature of issues and an interest

10 The concept 'basic quality' is explained in the Policy Document on Land Use. The government sees basic quality as 'the lower limit for all spatial plans' (VROM et al. 2004; 28), whereby statutory frameworks are created through standards for health, the environment, safety and security. The Quality Perspectives often refer to 'basic quality', usually in the form of limiting values for environmental ambitions (see Chapters 7 and 8).

in approaches that regarded environmental interests as part of a comprehensive policy for the physical environment in which different aspects were balanced against each other. New terms were introduced into policy, and it was not always clear how they were to be operationalized.

In short, thirty to forty years of environmental policy has produced a colourful mosaic of approaches and policy lines. The first Methods for Environmental Externalities were developed as Dutch environmental policy became firmly established. In the mid-1980s, the Association of Netherlands Municipalities (VNG) launched its ‘Green Book’, which at the time was the ultimate example of the classic standard-based approach. It was also one of the first Methods for Environmental Externalities. But, as we have seen in this chapter, policy continued to evolve. New ideas led to new approaches and lines in environmental policy. Policy concepts were formulated and implemented. The exercise with Integrated Environmental Zoning produced some interesting insights. The reliance on standards and their compulsory nature confirm how the VS-IMZ fits in to the period of the traditional standards-based system. Quite different were the intentions of the ROM approach, which was developed at more or less the same time. Whereas Integrated Environmental Zoning was deficient in terms of an integrated and area-specific approach, the ROM designated-areas policy – despite having just as many limitations (De Roo 2001) – *was* considered a step in the right direction.

Although it is difficult to argue a direct causal link, the success of the area-specific approach in the ROM designated-areas policy undoubtedly contributed to the evolution of Methods for Environmental Externalities that focus on the specific aspects of issues experienced at a local and regional level. In the second half of the 1990s in particular, there was an almost explosive increase in the number of Methods for Environmental Externalities (for a survey, see VROM 2002). In some respects, the methods can be seen as a reflection of the environmental policy of the time. The relationship appears so close that it can be said that there was a method for environmental externalities for almost every environmental policy line. This is an interesting observation, especially given that the development of Methods for Environmental Externalities are still continuing. In the next chapter we present a comparative framework, which – based on the various lines of Dutch environmental policy – can provide insight into a selection of Methods for Environmental Externalities.

Chapter 3

Diversity Uncovered: A Comparative Framework for Methods for Environmental Externalities

A medium-sized municipal authority in the Netherlands is considering a reorientation of its policy for the physical living environment. The policy it has followed so far has been reasonably successful, yet there is a feeling that this policy is paying too little attention to cohesion and integration. The various departments are taking too little account of each other and as a result the policy is sometimes implemented inefficiently and with little effect. An environmental officer working for this municipal authority believes it is important to be involved at an early stage in thinking about and discussing the course of policy in the years to come. In the first discussions he points out the possibilities for change, but it seems as if he is not getting through to his colleagues in the Spatial Planning department. Despite everyone's input, the discussion does not go smoothly. Afterwards, on the way back to his office, he thinks about what happened. Where did things go wrong?

It is quite likely that this fictitious situation sounds familiar to you. A great deal has been written in recent years about the difficult relationship between spatial planning and environment health and hygiene. Yet even today, despite the many improvements that have been made, situations like this still arise. Generally speaking, someone in the Environment department will see environmental issues through different eyes than his colleagues in the Spatial Planning department. The spatial planner is looking for opportunities to develop the physical environment. In such situations, environmental policies are often considered as inconvenient constraints, especially if environmental considerations are looked at late in the planning process.

An interesting question is how local and regional authorities can deal with this problem. There are a number of options. This book explains the potential of methods for environmental externalities. These are methods designed to help build a bridge between spatial planning and the environment. Many such methods have been developed over the years. Each method approaches issues in its own specific way with its own specific objectives, and produces it own results. For example, there are methods designed for determining environmental quality in a particular area. There are also methods for enhancing and monitoring the contribution of environmental considerations to spatial planning processes. Looking at the range of methods, it would seem legitimate to conclude that there is an appropriate method available for every type of issue.

This wide variation allows various methods to be chosen for various situations. Hence, method A is used for one particular issue, but for another issue it is better to leave method A in the drawer and use method B. But what is the difference between methods A and B, and what are the criteria for choosing between them? These are not easy questions. In order to make a well-informed choice, it is necessary to have an understanding of the principles, characteristics and mechanisms of each method. In this chapter we present a comparative framework that can be used to distinguish between the selected methods for environmental externalities. The framework has three components. The first component can be used to determine the *position* of the environmental-externality method in relation to Dutch environmental policy (Section 3.1). The second component combines five elementary *characteristics* of the method in a way that clarifies their main differences (Section 3.2). The third component of the comparative framework involves looking at the applicability of the methods in the different *phases* of a spatial planning process. The framework is summarized in Section 3.4.

3.1 The Methods in Relation to Environmental Policy

The discussion in the previous chapter on developments in Dutch environmental policy is one of the main steps towards establishing the first component of the comparative framework. The discussion showed that, in the past thirty to forty years, environmental policy has produced several different approaches. The traditional standards based nature of policy was abandoned and replaced with an approach whereby environmental interests are balanced against other interests. There was also a shift away from environmental policy dominated by central-government interventions towards environmental policies that are part of comprehensive local policy for the physical living environment. In other words, Dutch environmental policy is being decentralized.

This observation provides the first 'handles' for developing the comparative framework envisioned here. In order to understand these, it is necessary to discuss the decentralization process. In general terms, decentralization can be regarded as 'a process, the aim of which is to transfer tasks and powers from a higher to a lower echelon in an organization, whereby the lower echelon both performs the task and assumes responsibility for it' (Elzinga and Hagelstein 1998; 111). In the case of Dutch environmental policy, the decentralization process is visible in the shift from a centrally governed framework-setting approach to a more flexible approach with scope for local solutions and more responsibilities for local authorities (for a discussion see De Roo 2001, Van Tatenhove & Goverde 2000). A decision to decentralize policy can be based on several considerations. Fleurke et al. (1997) name tailor-made solutions as an argument for decentralization. De Roo (2001) argues that authorities at lower levels of government are better able to make policy geared towards specific local circumstances. This consideration is the same thing as aiming for tailor-made solutions. According to Oosting (1984), decentralization is an essential precondition for democratic public administration. According to Kamphorst (2006), this is similar to the considerations underlying participative forms of planning, which can also be related to forms of planning in which government influence is reduced in relation to the role of local stakeholders and the general public (see also Van Tatenhove and Leroy 2001). It should be no surprise that many of the methods developed recently are based on the same participative principles (see Chapter 4 and Part B).

This brief survey shows that policy decentralization can be based on a number of different considerations. To summarize, the aim of decentralizing environmental policy is to create a better balance and combination of various policy objectives based on the specific local characteristics. This then produces more tailor-made policies geared towards the allocation of various land uses and activities in the local sphere. One consequence of this is a shift in the way in which traditional condition-setting and statutory environmental standards are used within such policy approaches. This has important consequences for the way in which methods for environmental externalities are applied.

In the discussion above, the decentralization of environmental policy in the Netherlands is explained in terms of a shift from a centralized top-down framework-setting approach to an approach geared to local circumstances. Decentralization can therefore be regarded as a 'shift in the approach to policy issues' (De Roo 2004; 23). The status, character and consequences of environmental ambitions and targets will depend on the degree of decentralization. This in turn will determine which type of method for environmental externalities is considered useful. In other words, methods for environmental externalities can be differentiated on the basis of the degree to which they rely on the direct application of central policy ambitions or the application of these ambitions within a more flexible decentralized and area-specific approach.

In some respects, methods for environmental externalities have thus evolved in parallel with the shifts in Dutch environmental policy. Hence the environmental zoning methods of the 1980s are normative and framework-setting in character – a description that can also be applied to the environmental policy of that period (see for example De Roo 2001). Later methods reflect the increasing emphasis on an area-specific approach that is more participative.

This book uses the decision-making model proposed by De Roo (2004), so as to clarify policy and policy decentralization (Figure 3.1). The first component of the comparative framework for methods for environmental externalities is based on that model. The decision-making model consists of three axes – the 'dimensions of planning and decision-making' (De Roo 2001, Hidding and Kerstens 2001). These dimensions are marked by extreme positions, as shown in Figure 3.1.

The position in the top left of the framework represents centralized governance in environmental policy, with straightforward fixed objectives. These points of departure are in line with the environmental policy of the 1980s and early 1990s.

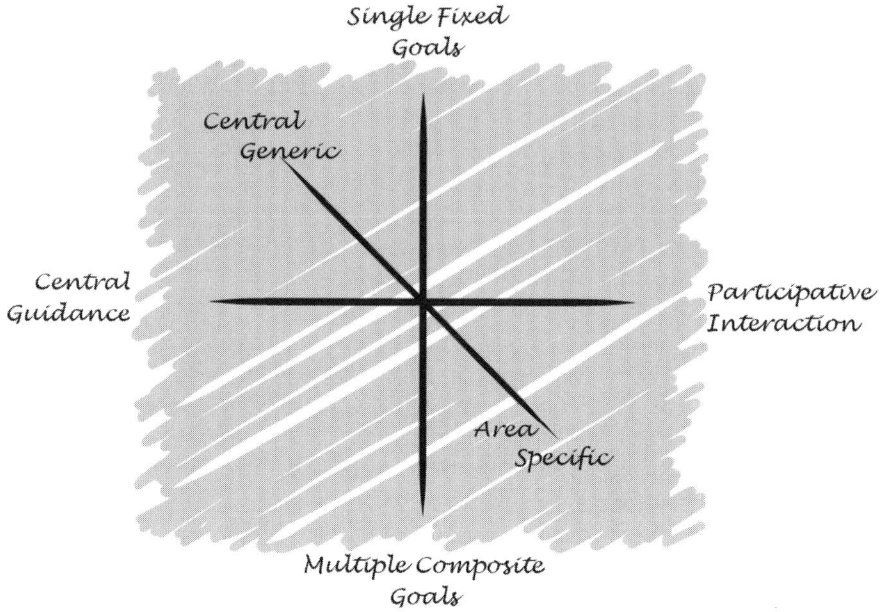

Figure 3.1 A decision-making model for decentralization (De Roo 2004)

The position in the bottom left of the framework represents a form of policy based on participative interaction and multiple composite and interdependent objectives. These are the extreme positions on the diagonal spectrum: at one extreme we have centralized generic policy, and at the other extreme we have area-specific policy. The process of decentralization is expressed in a diagonal shift from top-left to bottom-right (Figure 3.2).

Each expression of policy can be shown in the model in this way. The model therefore provides a basis for comparing different forms of policy. Similarly, methods for environmental externalities can be positioned in the framework. In a similar way to various policy approaches, methods for environmental externalities can also involve a centrally governed process or a more communication-oriented approach. The decision-making model can therefore be used to compare selected methods for environmental externalities.

In Part B of the book, each selected method for environmental externalities is discussed separately. We also determine the position of the methods within the decision-making model, based on their principles. These positions will be amalgamated in Chapter 11. A trend will then become evident during this process. The 'traditional' environmental-assessment methods will be positioned in the part of the framework that corresponds to centralized generic policy, while the methods developed more recently share more characteristics with decentralized policy. The resulting insights produce a basis for choosing between various methods for

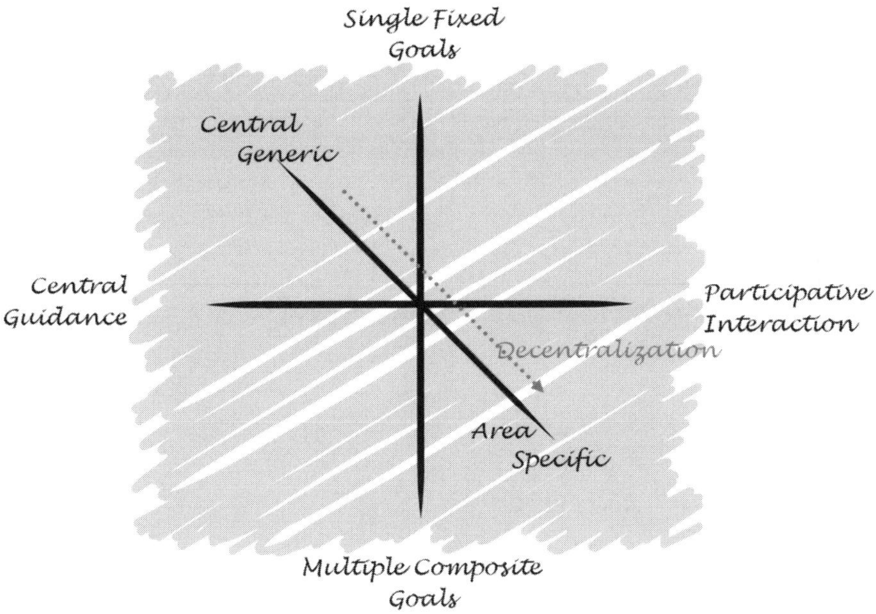

Figure 3.2 The process of decentralization shown in the decision-making model

environmental externalities, in accordance with the physical and organizational characteristics of the issue in question.

3.2 In-Depth: Characteristics of the Methods

In the first place, using the comparative framework will result in an understanding of the character of methods for environmental externalities. It produces a general picture whereby the character of the method is expressed in terms of either a generic and framework-setting approach or an area-specific and participative approach. In the second place, the comparative framework is used in an in-depth analysis of the method. The selected methods are assessed against five criteria in order to determine their character. The same criteria are applicable for each method and thus facilitate the comparison of the methods for environmental externalities.

The second application of the comparative framework is based on the 'IBO scheme'. IBO is the Dutch acronym for 'analysis instrument for policy development' (Borst et al. 1995). This analysis model was developed in the mid-1990s in order to gain insight into the policy instruments that can be used to resolve environmental/spatial conflicts. The IBO model has been used to analyze IMZ policy instruments, for example (Borst et al. 1995).

The core of the IBO model is a 'cake' consisting of three concentric rings that can be cut into 'wedges'. Each wedge represents a policy theme (criterion) for analyzing the instrument in question. Since the IBO 'cake' is made up of three concentric rings, each wedge also has three segments (Figure 3.3). A meaning can be assigned to each ring. In this case, the rings represent shifts in environmental policy, expressed in terms of centralized governance, decentralized governance and an intermediate form of governance. The inner ring represents centralized governance and the outer ring represents decentralized governance. A distinction is then made between the chosen themes (the 'wedges' of the cake) on the basis of these forms of government. The characteristics represented in each of the 'wedges' are explained below.

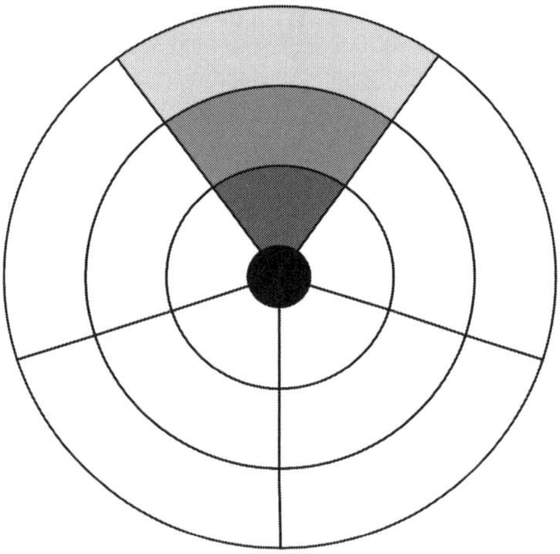

Figure 3.3 The basic form of the IBO model

Intention

The first characteristic to be considered is *intention* (Figure 3.4). This refers to the envisioned outcome that the application of the method produces. For example, VS-IMZ is based on a system of standards that produces a generic outcome. The environmental zone that is the outcome of the method is intended to be interpreted as a rigid framework that dictates the possibilities for spatial development in a given area. It thus represents a centralistic approach, based on generic standards. At the other end of the policy spectrum we have the 'Environment in Spatial Planning' method (MIRUP, *Milieu in Ruimtelijke Ordening*). This method is designed to help interpret environmental policy within the local and specific circumstances. The outcome of the method is intended to be used as a set of guidelines or a source of inspiration, not as a 'final' or 'definite' answer. Intermediate forms also exist, for example in expressing outcomes not as a generic or solely indicative outcome, but as a 'bandwidth' of remaining policy options.

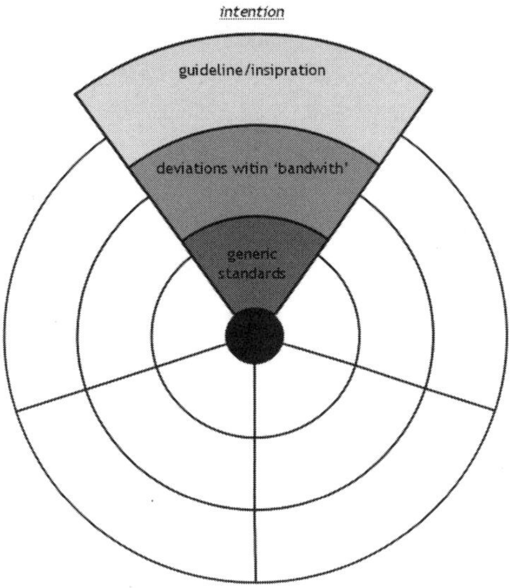

Figure 3.4 Intention as a thematic perspective in the comparative framework

Integration

The coordination and integration of spatial planning and the environment has been a subject of discussion for many years within all policy fields relating to the physical living environment (see for example Bakker et al. 1998). In the above discussion on environmental policy in the Netherlands, we have identified various periods that have aimed for some form of integration. Hence, this theme is also relevant for methods for environmental externalities. Some methods focus only on representing the various forms of environmental stress, reflecting the ambitions of internal integration in which various environmental themes must all be represented and combined. Other methods address other key aspects influencing the quality of the physical environment, in addition to environmental aspects. These are therefore intended to represent this quality, based on the ambitions of external integration where environmental, spatial, social and other priorities are to be taken into account together (Figure 3.5).

Pattern of Interaction

In recent decades, the way in which the various actors are involved in planning processes has changed drastically. In the early period of Dutch environmental

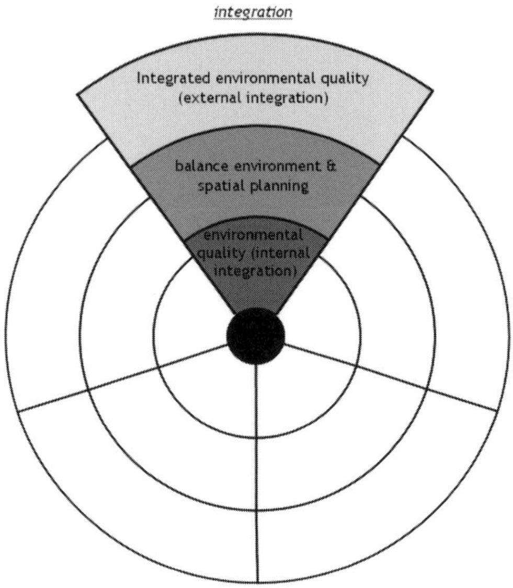

Figure 3.5 Integration as a thematic perspective in the comparative framework

policy, it was primarily the central government that set the frameworks within which the lower levels of government could implement policy. Today, the government is only considered as one of several actors involved in policy for the physical environment. This change also applies to methods for environmental externalities. On the one hand there are methods designed to be used solely by local authorities, and on the other hand there are methods whereby local authorities are supposed to work with other stakeholders in order to find a solution to spatial-planning issues (participation). VS-IMZ is an example of a method in which the government authority is the dominant actor. In the case of the MIRUP method, the government is regarded as only one of several actors involved (Figure 3.6).

Differentiation

In recent years, emphasis on the area-specific characteristics of issues has grown. This is reflected in the many area-specific approaches that have evolved (for a discussion see De Roo and Schwatz 2001, Oosterhof et al. 2001). This type of approach is also reflected in a number of methods for environmental externalities, in particular that of Quality Perspectives. In these methods, local and specific characteristics form the basis for finding solutions to spatial-planning issues (Figure 3.7).

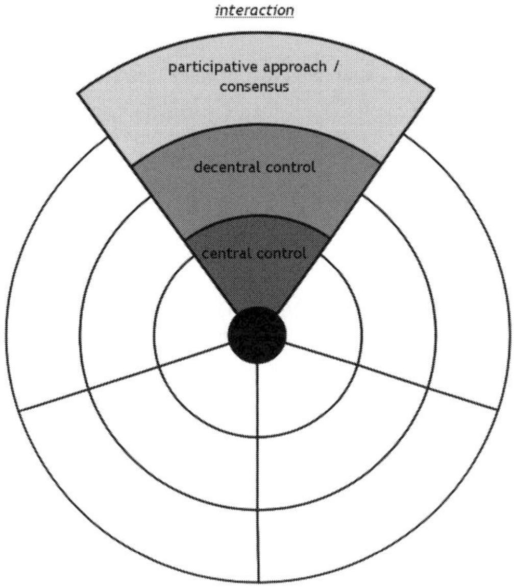

Figure 3.6 Pattern of interaction as a thematic perspective in the comparative framework

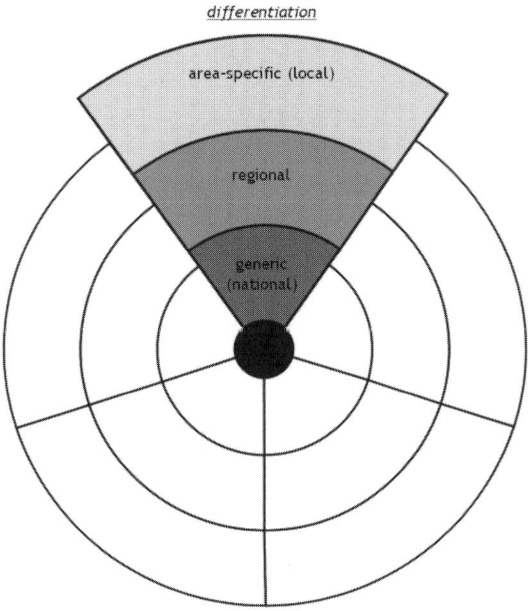

Figure 3.7 Differentiation as a thematic perspective in the comparative framework

Conformance and Performance

Methods for environmental externalities can be applied in various ways. This depends partly on the structure of the method and partly on the party implementing the method. The characteristic *conformance & performance* relates to how the results of a method are supposed to influence the eventual policy for the living environment (De Lange 1995). Again, VS-IMZ serves as an illustration. The aim of this methodology is to establish administrative environmental zones around clusters of activities with an environmental impact. The effect of VS-IMZ is therefore – in terms of the effect envisioned – a very direct one. This is less true of other methods. The Quality Perspective method referred to above usually results in a vision for the environment that serves as input for the spatial-planning process. Here there is also a supposed 'effect', albeit one of a less framework-setting nature (Figure 3.8).

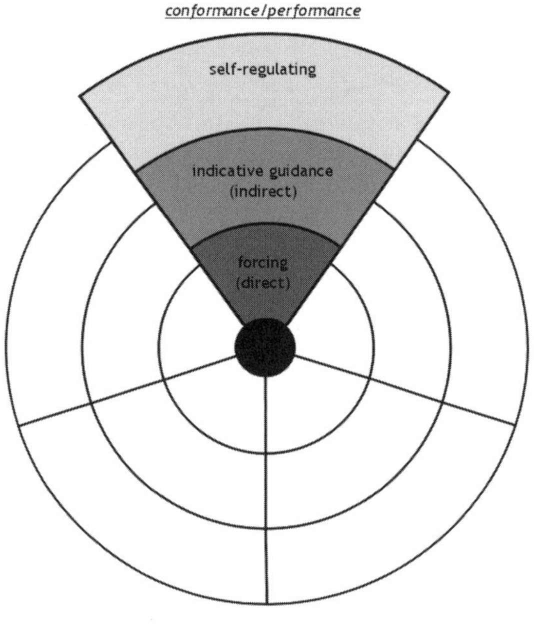

Figure 3.8 Conformance and performance as a thematic perspective in the comparative framework

The identification of the five thematic perspectives completes the second part of the comparative framework (see Figure 3.9). The inner ring of the IBO cake represents segments that characterize centralized generic policy (see also Figure 3.2). By contrast, the outer ring represents segments that are assumed to be representative of a decentralized policy environment, which is characterized

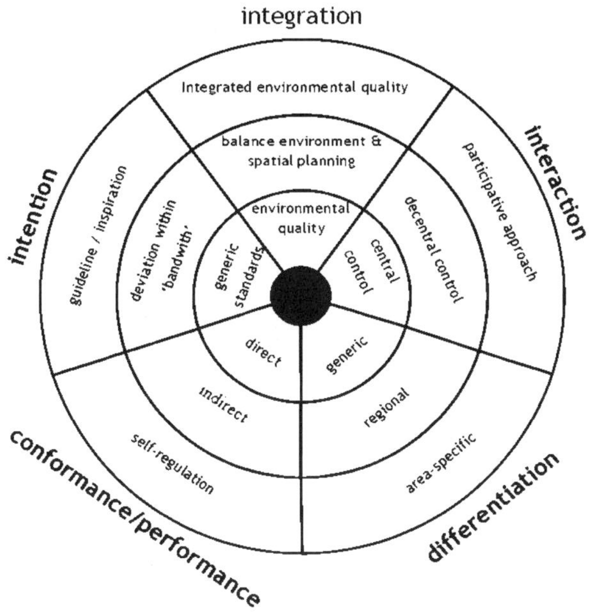

Figure 3.9 Thematic perspectives in a comparative framework

by an area-specific approach and participative orientation. In Part B, the methods are entered in the appropriate segments. In Chapter 11, the methods are compared.

3.3 Positioning of Methods in Planning Processes

The part of the comparative framework for methods for environmental externalities relates to the scope of applicability of the methods in spatial-planning processes. Some methods, for example, are designed to *provide information* at an early stage in the process. Other methods can be used to *develop concrete solutions* for the issue at hand. Still other methods can be used for *evaluation purposes*. These methods may be designed to compare alternative policy options. There are also methods that provide 'handles' for the entire planning process (*process-supporting methods*). This third part of the comparative framework is intended to indicate loosely when certain methods are more or less applicable during various phases of the planning process.

A spatial planning process can be divided into various phases or stages. This can done, for example, by specifying how various actions or steps to be taken are essentially indicators that a new phase has started (for overviews, see De Roo and Voogd 2004, Hoogerwerf and Herweijer 1998, Kickert 1986). A planning process can also be segmented more generally, as in the comparative framework presented here. The planning process is assumed here to consist of five main phases (Figure 3.10): initiation, definition, design, decision-making and – finally – implementation and monitoring. Figure 3.10 shows a highly simplified planning process. In practice, phases may overlap, be omitted or take place simultaneously. The phases are explained below.

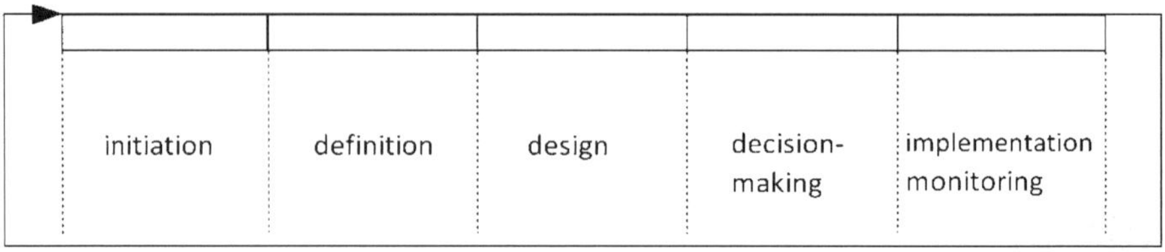

Figure 3.10 The five phases in the spatial planning process

The process begins with the *initiation phase*. In this phase there are not necessarily any concrete plans – these can be drawn up as the process progresses. However, there are grounds for initiating the process. During the first phase, the nature of the issue is established and the stakeholders and characteristics of the area identified, etc. The environmental sector does not initiate spatial plans, but it can make a valuable contribution by ensuring that the relevant environmental information is available in good time and that environmental health and hygiene conditions are identified. Environmental atlases are an example of an instrument for this phase.

The initiation phase can be regarded as an exploration of the issue and desirable solution directions. In the *definition phase*, the spatial issue is delineated: policy goals (planning actions) are formulated. This may involve, for example, stipulating the types of homes to be built or the facilities for the area in question. The desired environmental qualities are also formulated in this phase, in the form of environmental ambitions or otherwise. Quality Perspectives are instruments that can be used in this phase of the planning process.

Once policy goals have been determined, the *design phase* begins. During this phase, various methods for environmental externalities can be used to translate policy goals (environmental ambitions) into concrete measures for implementation in the final design. The chosen goals will determine which measures are required. Some methods for environmental externalities have a ‘catalogue of measures’ that can serve as a source of inspiration.

Choices can be made in the *decision-making* phase. In this phase it is important to clarify to what extent the draft plans comply with the ambitions that have been formulated. This can be quite simple if the environmental input in earlier phases has been straightforward (e.g. a programme of requirements). If this is not the case, a range of methods for environmental externalities can be used in this phase to weigh up the various alternatives in the light of the ambitions set.

The final phase is the *implementation and monitoring phase*. This phase is also important for the environmental sector. The ultimate goal is not simply to incorporate environmental ambitions in the design and definition of spatial projects. A further goal is to incorporate these environmental ambitions in the final result. This means that a degree of monitoring is required during implementation. Monitoring can also be important in terms of accountability to the local community or administration. It may also be necessary to make policy adjustments. In this phase, methods for environmental externalities are useful for providing further insight into the planning area.

3.4 Conclusion: The Comparative Framework in Focus

In this chapter we have described the comparative framework for methods for environmental externalities. The framework has three components. In the first place, it positions the methods in relation to Dutch environmental policy. Environmental policy has proved to be an evolving policy field, and methods for environmental externalities evolve in step with it. Gradually its focus has shifted to area-specific policies in which environmental ambitions are regarded as a true integrative part of local governance. It confirms how the external integration, most notably with spatial planning, currently manifests itself mostly on a local level. Hence, more recent methods are designed to facilitate this integration and form true bridges between environmental and spatial-planning policies. In the second place, the framework examines the methods in more detail. The methods are characterized using five criteria, which makes it easier to compare them. Finally, the framework assesses the scope of applicability of the methods in the spatial-planning process. The framework, a combination of Figures 3.1 and 3.9, is illustrated in Figure 3.11.

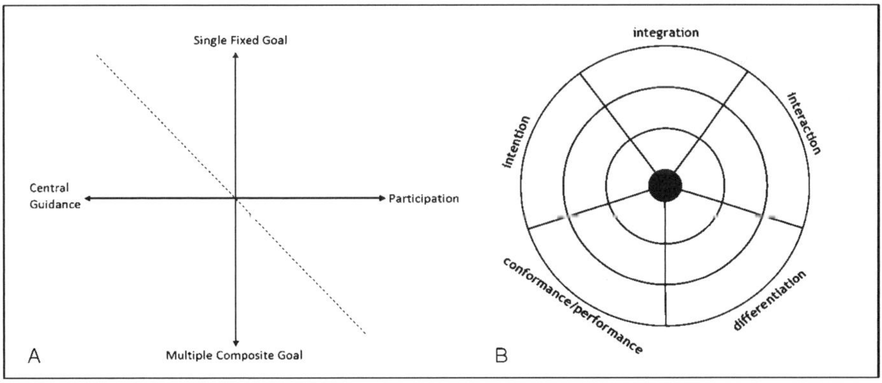

Figure 3.11 Comparative framework for methods for environmental externalities

In Part B of this book we apply the comparative framework to sixteen selected methods for environmental externalities. Each method is analyzed separately. This process results in the sixteen analysis diagrams as shown in Figure 3.11. Using these results, it is relatively easy to identify the differences between the methods. In order to retain the 'bigger picture', the methods are divided into categories. The categories are explained in the next chapter.

Chapter 4
Methods in Different Shapes and Sizes: Variation for Coping with Varying Circumstances

The purpose of methods for environmental externalities is to enhance understanding and support decision-making. In other words, they are about helping to solve spatial issues and ensuring that the environmental sector is involved in the planning process. Spatial issues may be concrete issues such as a restructuring project or preparations for a new residential area. But the matter may also be more general, such as formulating a vision for an area. Whatever the intended goal, it involves changing an existing situation into a more desirable situation. Due to the large number of methods for environmental externalities, this can be done in many different ways.

With the help of the VNG's well-known 'Green Book' (see Section 1.1, Chapters 2 and 5), quality standards are achieved through the enforcement of a minimum distance between environmentally harmful and environmentally sensitive activities. This technical approach is almost diametrically opposed to the way in which the more communication-oriented methods deal with such matters. For example, municipalities can use the LOGO guidance instrument to develop an integrated vision for an area. The LOGO method emphasizes cooperation with other stakeholders. These two methods (LOGO and the technical approach) can be seen as the extremes of a spectrum. At the points between the two extremes we have methods with different 'mixes' of the technical and communication-oriented approaches.

The desired outcome of the VNG method and LOGO is therefore the same: to change an existing situation into a more desirable situation. Nevertheless, in certain cases, one method will be more appropriate than the other. The VNG method results only in the basic separation of the source and recipient of the pollution. In general, this method will be appropriate only in relatively straightforward situations. The LOGO method approaches issues in a different way. Cooperation between the relevant policy sectors is seen as the key to successful area development. This means that, from the outset, all the relevant factors (including environmental factors) and stakeholders can be taken into account. Therefore LOGO is not limited to straightforward issues. In 2006, for example, the method was applied to the Stormeiland redevelopment project in an area where heavy industry has traditionally been based. The project proved problematic in terms of achieving a balance between the redevelopment of industrial sites and the development of combined residential and business districts in the project area. The LOGO method was used to formulate a vision for the area,[1] thereby demonstrating that the method is also suitable for more complex situations.

The choice of method will depend on a number of different factors. These may relate to what the stakeholders want to achieve or to the characteristics, background and complexity of the situation in question. In the second Part of this book we will discuss a selection of methods for environmental externalities that have been developed recently, as well as environmental assessment methods that have now become ‘classic’. The discussion will contain many examples of actual situations in which the methods have been used. This will lead not only to a greater understanding of how the methods work, but also of the results in practice. This information can assist in the process of weighing-up and selecting methods.

We will discuss sixteen methods for environmental externalities. In order not to lose sight of the ‘bigger picture’, the methods are divided into categories. This chapter gives a brief introduction to the categories and the methods placed in each category. It should be noted that some methods fit into more than one category. The methods have been categorized according to their most salient features. The position of each category in Dutch environmental policy is described on the basis of the framework presented in Chapter 3.

4.1 From Environmental Standards to Spatial Contours: Environmental Zoning Methods

The first methods for environmental externalities were developed in the 1980s. Examples are the VNG zoning method known as VS-IMZ (Provisional System for Integrated Environmental Zoning) and the Bubble Method (*Stolpmethode*) developed by the Amsterdam municipal authority. These methods were regarded as environmental assessment methods because of their quantitative and evaluation-based approach.

The zoning methods (Figure 4.1) account for a large proportion of this first generation of methods environmental externalities. The purpose of environmental zoning is to identify problems (and potential problems) between environmentally sensitive and environmentally harmful activities. A new industrial estate close to a suburban residential district can produce problems for various reasons. In the first place, there is a risk of environmental nuisance from industrial activity: smell, noise, risks and even light pollution are possible. Problems can also be expected with transport routes. The consequences of these problems can also be multifaceted. Excessive noise levels can have consequences for health (sleep problems, stress) and the attractiveness and desirability of a neighbourhood.

1 This will be discussed in Chapter 8, which deals with zip methods.

Figure 4.1 Environmental zoning methods in Dutch environmental policy

By means of standards, environmental zoning methods can clarify these problems and propose the separation of harmful and sensitive activities. This can be done in several ways. The VNG's 'Green Book' uses extensive tables of distances. A distance is set between almost every type of industrial activity and a peaceful residential area. Adjustment factors can be used to determine distances for other types of area. The VNG method is popular and widely used. However, it is not adequate for more complex situations, for example because it does not take account of accumulated environmental load and situation-specific factors. The VS-IMZ method *does* do this. Measurements and calculations are used to establish a 'tailor-made' environmental zone around a complex of activities that are harmful to the environment. This administrative zone imposes restrictions in terms of spatial development opportunities in the area in question and specifies acceptable levels of environmental load.

Depending on the situation, zoning methods can prove to be powerful, welcome instruments for reinforcing local environmental policy. The methods can be straightforwardly applied and need not be overly expensive. A disadvantage is that, due to their generic character, these straightforward methods will not be adequate for more complex issues because too many factors are involved. This does not apply to VS-IMZ, which was developed precisely for this purpose. But this method also has its limitations. The planning implications of the VS-IMZ method had such disastrous consequences for some cities that it was deemed unworkable – by the local authorities in particular (see for example De Roo 2001).

Despite this, VS-IMZ can be successfully used to determine the environmental load in an area, on the basis of actual and situation-specific assumptions.

Today, virtually no modern variants of zoning methods are developed. The concept of environmental zoning therefore seems to be outdated. Nevertheless, environmental zoning is still used as part of current environmental policy. This involves not only area-specific policy (see Chapter 2), but also target-group policy. A recent example is the measures relating to European emission standards for lorries. Lorries with a Euro 3 or Euro 4 classification, for example, must be fitted with a particulate filter if they are used in certain city centres.

The environmental zoning methods will be discussed in Chapter 5. The VNG method (Section 5.1) has now been updated three times, so the version currently in use is the fourth version. The VS-IMZ (Section 5.2) has never progressed beyond 'provisional' status. Nevertheless, as we shall see, this is an important method in terms of its structure, and it still has potential.

4.2 Between Quantity and Quality: Checklist Methods

It is a much-discussed subject: the involvement of environmental interests at an early stage in spatial planning processes. From the outset, this has proved difficult to achieve. The environmental sector is often involved too late in spatial planning processes, which means that the environment runs the risk of being overshadowed by other interests (e.g. economic interests). Furthermore, it is not always easy to form a clear picture of the health effects of certain interventions. Over the years, various methods have been developed with a view to ensuring that the relevant health and environmental aspects are taken into consideration at the appropriate stage in the planning process. These methods are known as *checklist methods* (Figure 4.2).

One of the first checklist methods was developed by the Amsterdam municipal authority, and is known as the Environmental Performance System (*Milieuprestatiesysteem*) (Section 6.1). The purpose of this method is to guarantee that new spatial development achieves predefined quality levels. Environmental measures are selected from a list in order to contribute to liveability in the area in question. The *Milieuplaberum* (planning & decision-making process for spatial measures) is another checklist method (Section 6.2). This method was also developed by the Amsterdam municipal authority and some aspects of it are similar to the Environmental Performance System. However, the latter is linked to a zoning/land-use plan, while the purpose of the *Milieuplaberum* is to integrate environmental demands in the spatial planning process at an early stage.[2] These two checklists focus mainly on environmental interests. This is also the case with

2 This means that *Milieuplaberum* could also be categorized as a zipmethod (Chapter 8, Section 8.4). However, the 'catalogue of measures' is such an important part of the method that *Milieuplaberum* has been categorized as a checklist method.

Figure 4.2 Checklist methods in Dutch environmental policy

the Health Effect Screening method (Section 6.3), but this is explicitly based on the health perspective. This checklist was developed by the GGD (Municipal Health Services) and must be used at the beginning of a planning process. Only then, so the reasoning goes, can the health effects of policy intentions be identified. If necessary, policy can then be adjusted.

Checklist methods can produce a great deal of information in a relatively short period of time. The question is then, for example, whether all the relevant environmental aspects have been covered. In the case of Health Effect Screening, this involves aspects relating to health. The checklist methods will be discussed in Chapter 6.

4.3 Towards Area-Specific Customized Solutions: Quality Profiles

The category 'quality profiles' comprises methods that have mainly been developed recently (Figure 4.3). An important characteristic of this type of method is that the environment is regarded as a factor that can be weighed against other factors. This is usually done on the basis of the features that characterize the area in question. This makes it possible to set different environmental-quality targets for industrial areas and suburban residential neighbourhoods or city centres.

Quality profiles can be used to find area-specific solutions for spatial projects. In general, the approach is communication-oriented. Close cooperation and consultation between stakeholders is important for the successful application of

Figure 4.3 Quality profiles in Dutch environmental policy

these methods. But there is one further condition for the successful application of quality profiles: they must be applied at an early stage in planning and decision-making processes. The explanatory notes to the methods also state that this must be done as early as the preparatory stage of the decision-making process. Environmental interests thus become a 'visible' part of the planning process and are set down in the form of quality ambitions that are the basis for the subsequent stages of the planning process. In the fourth National Environmental Policy Plan (NMP4), municipal authorities are required, among other things, to translate environmental ambitions into tangible measures for improving liveability. The quality profiles can assist in this process.

The application of quality profiles reflects the differences between areas. These are not necessarily only physical differences. In some methods, attention is also paid to social and economic aspects. In general, environmental aspects form the basis for the quality profile. In most cases, the first step is an extensive study of the area with a view to establishing what type of area it is. Area types can be seen as predefined sets of limiting values and target values. The area type is a practical basis for defining environmental ambitions. The area study also produces information on the environmental aspects and stakeholders that are relevant to the situation. An ambition is formulated for each environmental aspect. All the ambitions taken together form the environmental quality profile.

Quality profiles have now been in use for some time. Practical experience shows that the methods are a welcome supplement to local and regional environmental policy. This applies to situations in which the profiles are the only method used to find a solution to a spatial issue, as well as to situations in which the profiles are part of an environmental policy plan. Quality-profile methods are discussed in Chapter 7. The following methods will be dealt with: the 'Bandwidth' method (BBM, *Bestuurlijke Bandbreedte Methodiek*) developed by the Utrecht municipal authority (Section 7.1), the Environmental Quality Profiles developed by the Maastricht municipal authority (Section 7.2) and the neighbourhood-oriented environmental targets (*Wijkgerichte Milieustreefbeelden*) implemented by the Groningen municipal authority (Section 7.3).

4.4 'Handles' for Spatial Planning Processes: Zip Methods

In recent years, a new category of methods for environmental externalities has evolved. We are referring to these methods as zip methods (Figure 4.4). Their purpose is to increase the input of the environmental sector in spatial planning processes. Some of the methods in this category even present a structure for the spatial planning process, in which attention is consistently paid to environmental factors.

Figure 4.4 Zip methods in Dutch environmental policy

Zip methods are based on the principle that more than one road leads to Rome; in other words, there are several ways to improve the quality of the living environment. The structure presented offers stakeholders sufficient scope to achieve a balance between spatial planning and the environment in the solution for the spatial project in question. This involves using more than one method; these are instruments that comprise a set of processes and techniques.

An example of such methods and techniques is the 'layered approach' (see Chapter 8) for identifying and analysing the planning area and the spatial project. Another example is the use of quality profiles to formulate policy goals. Because the methodologies are expressly presented as a guidance instrument, there is scope to supplement them with existing methods. The instrument thus provides the framework within which the methods can be used for the spatial project. For example, an area study could be supplemented with Health Effect Screening. The VS-IMZ method could also – provided it was applied in a different way than that intended – provide sufficient insight into the environmental aspects that are relevant to an issue. The guidance instruments are hence described by their designers as 'toolboxes' for professionals involved in designing/shaping the physical environment.

Well-known guidance instruments include LOGO (Section 8.1) and MIRUP (Section 8.3). LOGO is the Dutch acronym for Local Area Typology and Quality of Life, and MIRUP is the Dutch acronym for Environment in Spatial Planning. We will also discuss another zip method known as MILO (Section 8.2). The method is based on the original MILO guidance instrument. It also uses components from LOGO and MIRUP. These methods will be discussed in Chapter 8.

4.5 Environmental Information as a Basis for Decision-Making: Information Methods

Information methods are a special category within the selection of methods for environmental externalities discussed in this book (Figure 4.5). All the other methods for environmental externalities discussed here operate according to an input-output model. The method is used to process information (input) in order to obtain a certain result (output). For example, an environmental quality profile is used to produce a target scenario for the living environment. An environmental zoning method results in an administrative zone within which spatial development is either subject to restrictions or not permitted. This is not the case with information methods.

Examples of such methods are environmental atlases, which are becoming increasingly popular. An atlas is used to present information relating to the environment. They can serve many purposes. Environmental atlases can be used to improve internal coordination between departments. They also make it possible to inform the local community about the quality of the living environment in an active way. In fact, recent statutory provisions have made this compulsory for local authorities. A third way in which environmental atlases can be used is as preparation for the application of other methods for environmental externalities.

The environmental-quality profile method, for example, requires a great deal of information. Traditionally at the beginning of a spatial planning process, the current environmental quality first needs to be analysed, and then targets can

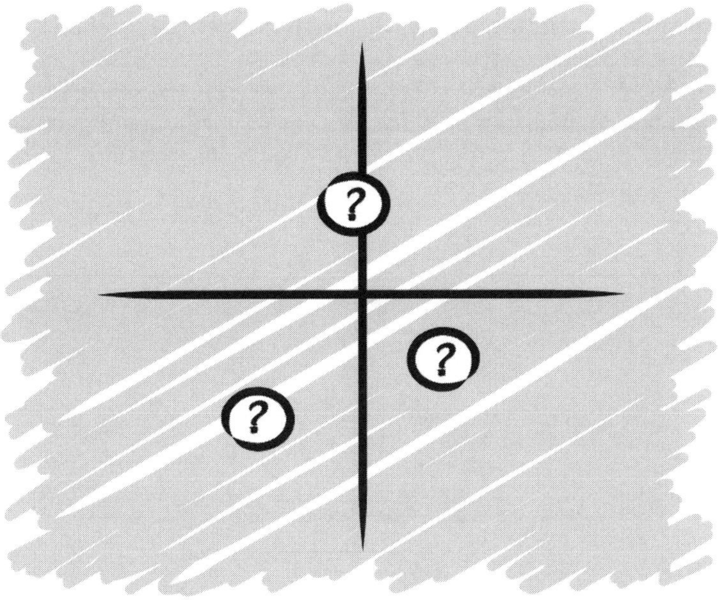

Figure 4.5 Information methods in Dutch environmental policy

be considered. Decision-making processes can be accelerated if the required information is available as standard to local authorities. In many cases, some of this information is already available, albeit in a fragmented and unstructured form. An environmental atlas can assist this process. The relevant environmental information is gathered and provided in a structured form, thereby ensuring that it is available at the beginning of the process.

The 'information methods' category will be discussed after the other categories. There is a good reason for this. A classic environmental-zoning method (VS-IMZ) will be discussed in Chapter 5. As the discussion will show, this method can serve as an information method in certain situations. The method's consequences for planning currently stand in the way of this. The VS-IMZ method and environmental atlases can therefore be regarded as extremes of a spectrum. At one end we have the centralized framework-setting approach of zoning methods, and at the other end we have the information methods, which are communication-oriented and take account of local considerations. In Chapter 12 we discuss the significance of these observations for the application of methods for environmental externalities as part of a decentralized environmental policy.

Chapter 10 examines two information methods. The first of these is the Environmental Atlas of the Deventer municipal authority (Section 9.1). The atlas has been in use for several years. The second information method is the Groningen Environmental Atlas (Section 9.2).

4.6 From Thought Models to Design Models: Other Methods

There will always be methods that do not fit into the categories defined here, which means that such methods can be placed anywhere within the framework (Figure 4.6). This applies to three of the sixteen selected methods, which have been put into the 'other methods' category.

Figure 4.6 Other methods in Dutch environmental policy

These methods have been put into this category because they are too dissimilar to the other methods. They also have a unique character, which makes them sufficiently inspiring to be mentioned here. The Bubble Method (*Stolpmethode*) developed by the Amsterdam municipal authority is conceptually very interesting and unusual, and therefore does not fit into any of the other categories. The aim of the method is to reduce the total environmental load in an area, city or region. In order to achieve this, the concept of 'compensation' between various types of environmental pollution was introduced. A second method in this category is the ROMBO tactic (Section 10.2), which emphasizes communication and interaction with all stakeholders in a spatial planning process, whereby the environmental theme seems to be incidental rather than essential. The ROMBO method was developed in order to introduce sustainability aspects into planning and decision-making processes. The final method in this category is the Environment Maximization Method (*Milieu Maximilisatie Methode*) (Section 10.1). This is a design method for formulating solutions to spatial projects. It also uses environmental themes that are in some way relevant to sustainable spatial development.

One of these methods, namely the Bubble Method, has never been used in practice. Rather than being a practical policy instrument, it is primarily based on a 'balance' approach for dealing with complex urban problems. The same can be said of the ROMBO tactic. The striking thing about this method is its extreme position on the technical-communicative spectrum. Nevertheless, these three methods are certainly of value in the debate on policy theory (including environmental policy). These methods will be discussed in Chapter 10.

4.7 Conclusion

The categorization used here is useful for several reasons. In the first place, it makes it easier to compare the different methods in terms of their usage and aims. For example, the reflective nature of the Environmental Atlas is not really comparable to a checklist method such as the Amsterdam Environmental Performance System. For each category, we discuss methods for environmental externalities that are similar in terms of their starting points, objectives and mechanisms. In the second place, the categorization can assist the selection of an appropriate method for environmental externalities. In certain situations, it will be clear that a quality profile or zip method should be used. The use of categories limits the number of alternatives.

In the following chapters (Part B of the book), the methods for environmental externalities are discussed separately by category. All methods will be analysed in the same way (see Chapter 3 for an explanation of the analytical framework). Part C of the book deals with the main differences between the categories and with the main ways in which they can be compared.

PART B
Methods and Tools

Chapter 5
Environmental Zoning Approaches: Separating the 'Good' and the 'Bad'

By the mid-1980s, environmental zoning had become a reasonably common phenomenon in environmental policy. In different parts of the Netherlands, spatial developments were already limited by the use of environmental zones. In the Wadden Sea area and the Veluwe, for example, spatial developments were no longer automatically permitted. These are areas where ecological values play a central role. However, from the viewpoint of environmental nuisance, environmental zones were also being used. The Rhine Estuary area and the area around Schiphol airport are examples of this.

Until the mid-1980s, however, there was no straightforward methodology available to determine the desired distance between an environmentally sensitive function and an activity with environmental impact. Some municipalities did have their own approach, for example Spijkenisse, Amsterdam and Maastricht, but there was no question of coherence or unified ways of working. They were independent initiatives. This changed with the publication of 'Businesses and environmental zoning' (1986; 'Bedrijven en milieuzonering') by the VNG (Association of Dutch Municipalities). In the years that followed the publication of the 'Green Book', as the approach quickly came to be known, it would evolve into the standard methodology to be used by environmental and spatial planners. Even today, the 'Green Book' is widely used.

The success of the VNG's zoning approach did not pass unnoticed. New approaches quickly began to appear, some based on the VNG approach, some not. Examples include the Zwolle Environmental Assessment Approach, the Lelystad Zoning Approach (winner of the National Prize for the Environment in 1991) and the Groningen Environmental Assessment Approach. Towards the end of the 1980s, the Ministry of Housing, Spatial Planning and the Environment (VROM) also responded. In the wake of all these separate initiatives, it decided to develop a system for integral environmental zoning, which was also to become part of the Environmental Management Act. The discussion on the principle of environmental zoning was now well under way.

Characteristic of environmental zoning approaches is the way that environmental problems are tackled; i.e. through drawing up frameworks and presenting technical arguments. To this end, the maximum tolerated level of environmental stress was translated into environmental standards. These environmental standards were translated into direct spatial planning consequences, such as ensuring that a specified certain distance was maintained between the environmentally intrusive and the environmentally sensitive activities. Hence, environmental zoning approaches are dictated by a policy focus on harsh frameworks and by their use of direct governance.

This chapter will discuss two environmental zoning approaches, both used widely at national level. They are the VNG approach (Section 5.1) and the system developed by VROM for integrated environmental zoning (VS-IMZ) (Section 5.2). Section 5.3 will then compare the two environmental-zoning approaches.

5.1 'Businesses and Environmental Zoning' (VNG Approach)

The history of the 'Businesses and environmental zoning' approach begins in 1986, the year in which the VNG published the first version of the approach. This approach quickly became known as the VNG approach, or the 'Green Book'. It soon turned out that the approach was very helpful in drawing up municipal zoning plans: based on the type of environmental stress, a distance table (see Table 5.1) could be used to find the desired distance from sensitive activities. The simplicity of the approach and the low user costs contributed to it becoming very popular very swiftly.

The first years of the VNG approach can be regarded as an experiment with a system for environmental zoning. Never before had there been a simple methodology available for implementing environmental zoning in practice. During this period there were many discussions about the principle of environmental zoning. These discussions, and the first experiences with the approach, were taken into account when it was updated.[1] The second version was published in 1992, again by the VNG. Seven years later, as a result of shifts in environmental policy, it was again time for a new version. Alongside new practical experience, developments in Dutch environmental policy played an important role in the creation of the third version.

This time, the most important motive for renewing the manual was the increasing interest in the situation-specific aspects of spatial problems. These aspects undermined the applicability of the VNG approach. Contributing to this problem were the universal distance tables in the approach, which were only suitable for determining the average desired distances between environmentally intrusive activities and a sensitive (residential) activity. In the centre of a city, however, noise is experienced differently than in a quiet suburb. In an agricultural area or an area rich in nature the experience is different again. The second version of the approach provided no means of adjusting for such differences in experience. The third version of the VNG approach that was published in 1999 did include the

1 When the distance tables were updated, less relevant themes and aspects from the original version were reconsidered. This resulted, for example, in matters such as vermin being omitted from later versions of the manual.

possibility of taking into account different types of area and, hence, differences in experience.

The aim of the current updated third version of the 'Businesses and environmental zoning' approach has remained 'to offer help with the responsible incorporation of economic activity in its physical environment' (VNG 1999; 10). Because it takes the environmentally intrusive activity as its starting point, the VNG approach can be typified as an *inside-out zoning approach*. This means that a 'shell' or 'zone' was created around the intrusive activity and, within that zone, spatial developments were subject to restrictions. Using this 'zoning' principle, the 'incorporation of the economic activity in the physical environment' was achieved in the first place by allocating indicative distances to different forms of environmental stress caused by virtually all known types of business and institutions in the Netherlands. The result of this exercise was the distance tables that, now updated, were included in the earlier versions too. Second, attention was paid to sensitive activities in and around the businesses and institutions so as to address the question of how to deal with the differences between areas and experiences of the various types of environmental stress produced by these business and institutions (area differentiation).

Procedure

The distance tables form the core of the VNG approach (see Table 5.1). These tables show which distances between businesses and residential areas are considered desirable. The distances are determined for different forms of environmental stress (e.g. odour and noise) and are based on previous experience and legislation. These distances are based on average business types derived from the Standard Industrial Classification (SBI; *Standaardbedrijfsindeling*) of Statistics Netherlands (CBS). The distances are therefore intended to be regarded as indicative when used in practice, in order to be sensitive to the characteristics of individual businesses.

The distance tables include various types of data. First, indicative distances are given for spatially relevant components of environmental stress (odour, dust, noise and danger). When selecting these indicative distances, the reference situation is a situation with average new businesses and a quiet residential area with little traffic. Second, the table includes indices for aspects such as traffic-attracting effects and visual nuisance. Each index extends from 1 to 3. Index 1 means that there are 'potentially no emissions or low emissions'. Index 2 means 'potentially significant emissions' and index 3 means that there are 'potentially very serious emissions'. Third, there is a division into environmental categories. These can be regarded as a simplified version of the indicative distances: the greatest distance to be maintained determines the category to which the business is allocated. This means that the maximum permissible distance can be seen at a glance. Finally, there is an indication of whether a type of business has an increased chance of soil pollution (B), air pollution (L) or various emissions (D).

Table 5.1 Partial reproduction of the distance table for business types (after VNG 1999)

Description	Distances					Indices		Greatest dist.	Category	Comments
	Odour	Dust	Noise		Danger	Traffic	Visual			
AGRICULTURE AND AGRICULTURAL SERVICES										
Horticulture: 1 – commercial buildings	10	30	30	C	10	1	1	30	2	BL
2 – unheated greenhouses	10	10	30	C	10	1	1	30	2	BL
3 – gas-heated greenhouses	10	10	30	C	10	1	1	30	2	BL
4 – mushroom farms	30	10	30	C	30	1	1	30	2	B
5 – mushroom farms with manure fermentation	100	10	30	C	30	1	1	100	3	B
6 – bulb-drying and preparation factory	30	—	—	C	10	1	1	30	2	B

The third version of the approach offered the possibility of adjusting the indicative distances for different types of area, based on the estimated sensitivity of an area. To help the user to do this, the manual therefore includes an overview of the sensitivity of nine types of area, taking a quiet residential area as the reference area. Where an exclamation mark (!) appears in a row of the table, the area is considered more sensitive to the type of environmental stress relating to the row as compared to the reference area. Similarly, if a minus sign (-) appears, the area is considered less sensitive. In the latter case, the minimum distance could be reduced by one step (see Table 5.2).

Table 5.2 Distance steps and environmental categories (after VNG 1999)

Step	**1**	**2**	**3**	**4**	**5**	**6**	**7**	**8**	**9**	**10**
Distance	10	30	50	100	200	300	500	700	1000	1500
Category	1	2	3.1	3.2	4.1	4.2	5.1	5.2	5.3	6

An example can help us to better understand how the approach works in practice. Let us take the example of an industrial estate that is situated near a rural area with a residential function. Such a residential area is an area type with possible correction steps. The industry in the example is a mushroom farm with manure fermentation (listed in Table 5.1). With regard to the environmental aspect 'odour', there must be at least one hundred metres between this farm and a quiet residential area. The noise aspect requires thirty metres. According to the sensitivity table in the approach (not illustrated), this rural area with a residential function is less sensitive to odour, which means that a shorter distance can apply to the odour aspect. In the meantime, such a rural area with a residential function is typified as more sensitive than the baseline measurement with regard to the noise aspect. This means that a greater distance must be applied. The new distances can be determined using Table 5.2. Thus the minimum distance for odour can be reduced from a hundred metres to fifty metres (no more than one step may be corrected). With regard to the noise aspect, the distance must be increased from thirty to fifty metres. In this and similar ways, the distances for the other environmental aspects can also be determined. Finally, the greatest distance determines the environmental zone around the environmentally harmful activity.

Practice

The VNG approach has often been put into practice. Municipalities in particular have made use of the distance tables when determining or adjusting zoning plans. The approach has also proved its added value in practice; it is quick, easy and cost-effective. The experiences of municipalities have been incorporated in the later versions of the approach. The approach is also used when assessing licence applications.

Other examples of its use in practice can be seen in various decisions by the Dutch Council of State (the Dutch court of appeal for citizens against executive branch decisions). The distance tables in the approach are used to substantiate a decision or argument. Although the Council of State indicates in virtually every decision that the distances are indicative (see for example Council of State 2006), it is also of the opinion that any deviation from the distances must be substantiated (Council of State 1997). Officially, the VNG approach has no legal status and is not compulsory but, in the case law of the Council of State, the 'Businesses and environmental zoning' approach has been accepted as a standard methodology for environmental zoning.

Discussion

The discussion of the VNG approach is divided into three sections. The first section assesses the position of the approach in Dutch environmental policy, based on the framework discussed in Chapter 3. The second section of the discussion involves a more in-depth analysis. The five distinctive characteristics of methods for environmental externalities as discussed in Chapter 3 will all be examined. The last section will concentrate on the applicability of the approach in planning processes.

Although the current version of the VNG approach contains modern elements such as the possibility of applying area differentiation, it still echoes the environmental policy of the 1980s. The indicative distances, for example, are formulated using the standards of that period. This means that the approach is legalistic. There are also traces of a hierarchical approach. The centrally designed environmental standards are applied on a local level. As a result, the instrument should be projected in the upper left part of the framework (Figure 5.2a), which shows the position of the VNG approach in relation to Dutch environmental policy.

This position in the framework is confirmed when the VNG approach Is examined using the five characteristics of methods for environmental externalities as presented in Chapter 3. The aim of the approach is to separate environmentally intrusive and sensitive functions. The basis for the separation is formed by the indicative distances, which are a translation of stringent and generic environmental standards. Partly due to the case law of the Council of State, the approach has acquired a rather mandatory character when used. This mandatory character is somewhat nuanced by the possibility to deviate from indicative distances. However, even then, clear reasons should be given. The approach is therefore placed in the inner ring of the model for the characteristic *intention*.

The rather mandatory character and its wide use in practice have generated much confidence in the approach. Despite its indicative character, the approach is often used without sufficient prudence. Consequently, the indicative distances are simply relied on as being sufficient without further consideration, something which occurred, for example, in the municipality of Sittard-Geleen. Here, in 1998, an environmental officer told a group of environmental planning students, who were visiting the town, how policies to control and contain the risks and stress caused by the vast DSM chemicals complex were solely based on the VNG method.

A weak aspect of the VNG approach is the fact that various forms of environmental stress (e.g. noise, odour, safety risks) are not considered in relation to each other. In practice, the environmental stressor that dictates the greatest distance is determinative. The degree to which other environmental stressors contribute is not considered relevant. This certainly undermines the integrative nature of the method. Nevertheless, as the approach does lead to a merging of spatial planning and the environment, the methodology is placed in the middle ring when assessing it in terms of the characteristic *integration*.

The approach is used by municipalities for tasks such as preparing zoning-plan procedures. It limits interaction with parties other than governmental actors. With relation to the characteristic *interaction*, the approach is therefore placed in the innermost ring.

When it comes to *differentiation*, the VNG approach is more flexible. The third version of the VNG approach allows for differences in area sensitivity. The degree of flexibility is reduced, however, as adjustments cannot exceed one distance step. The approach is thus placed into the middle ring for the characteristic *differentiation.*

Finally, there is the characteristic *conformance & performance*. When applying the VNG approach there are very immediate effects – the indicative distances are included in the zoning plan. As a result, the method for the characteristic *conformance & performance* can placed in the inner ring of the model. Figure 5.2b depicts the relationships between the characteristics.

The applicability of the VNG approach is mainly determined by its objective: 'to offer help with the responsible incorporation of economic activity in its physical environment' (VNG 1999; 10). The approach can mainly be applied when preparing planning processes. In addition, the approach is also used when assessing licence applications. Figure 5.1 depicts the overall picture of this overview.

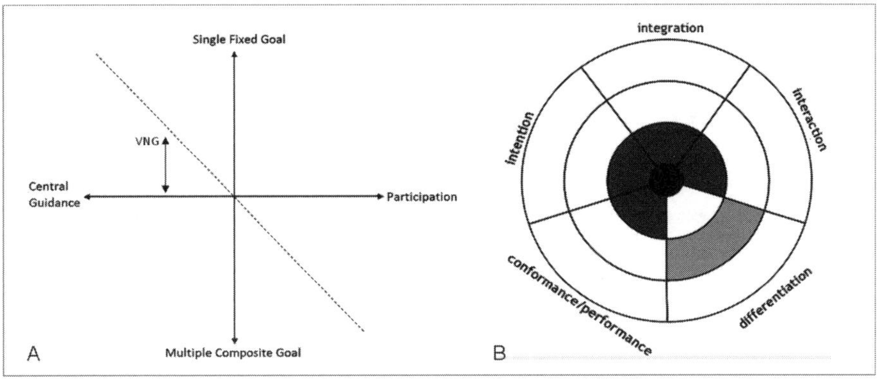

Figure 5.1 Characteristics of the Businesses and Environmental Zoning Approach

VNG Approach: Conclusions

The overview reveals various strengths and weaknesses. One of the main strengths of the approach is its simplicity. A relatively small amount of information is needed to be able to apply the approach. The second strength also relates to the simplicity of the approach. The desired information can be retrieved quickly. The approach is not labour-intensive and is therefore cost-effective to use. A third strength of the VNG approach lies in the years of practical experience gained with the approach. Consequently, it is generally accepted and is even regarded as the standard methodology in the case law of the Council of State. The final strong point is the completeness of the information included and the relative ease with which this can be translated into various types of area. As a result, the approach fits in with modern environmental policy.

A less strong point is the fact that it is based on indicative distances and average companies. Other distances may apply for companies that have taken measures to reduce emissions. However, the approach does not take this into account. Deviations from the distances – in line with the decision of the Council of State – must be substantiated. Another less strong aspect of the approach is its applicability. Due to its design, the approach is almost only suitable for new situations involving one type of environmental stress (see Figure 5.2). In practice, however, this rarely occurs. A third less strong point is the division into industrial categories. In practice it is possible for a certain company to fall into several industrial categories, or there is no appropriate category for the company. The approach does not allow for this. A fourth less strong point is the fact that the approach only takes account of the environmental stressor to which the greatest distance applies. This ignores the possible effects of cumulative environmental stress.

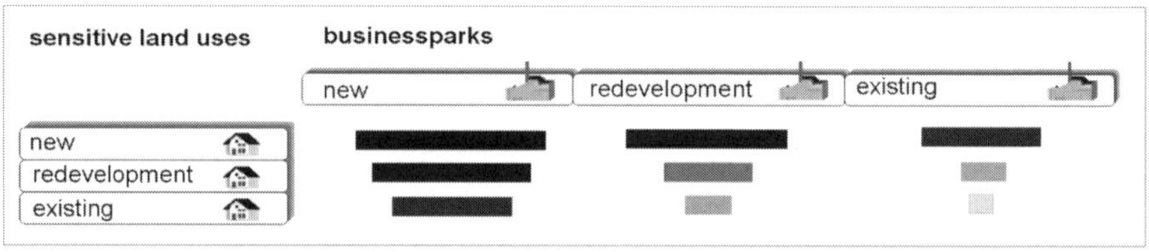

Figure 5.2 Decreasing applicability of environmental zoning (VNG 1999)

The simplicity of the approach makes it an interesting instrument for municipalities. Insight can quickly be gained into the desired distance between environmentally intrusive and sensitive functions. At the same time, however, this simplicity has also turned out to be the Achilles heel of the approach. The distances are indicative and thus not very detailed. This is necessarily so, because they are based on averages. As soon as a company has taken even minimal measures, the distances are no longer relevant. Its simplicity also means that the approach is less suitable for complex matters. And it is precisely this type of problem that is often encountered in the Netherlands. This means that the VNG approach does not apply to many of the current problems faced in Dutch environmental policy. However, despite these points, the fact remains that the approach can produce fast (thus cost-effective) and simple information about the environmental stress of virtually all types of industry in the Netherlands.

5.2 Provisional System for Integrated Environmental Zoning

The success of environmental zoning has not passed unnoticed. The VNG approach soon became a very popular tool for drawing up zoning plans. Quick, inexpensive insights into the relationship between environmentally sensitive and environmentally intrusive land-use functions were also possible with the VNG approach for smaller municipalities with limited resources and staff. Nevertheless, the approach had one major drawback. There were fewer possibilities in the event of complex situations with different types of environmental stress and with more than one source of nuisance. The need therefore arose for an approach that was able to address such situations. At the end of the 1980s, the Ministry of VROM therefore began to develop policy for what was called 'integrated environmental zoning'.

In the first National Environmental Policy Plan (NMP-1), the project 'Accumulation of sources and integrated environmental zoning' was presented (VROM 1989). At the time, this project was extremely innovative. Environmental problems were no longer divided into various categories of environmental impact. This is an important difference with respect to the VNG methodology. The ultimate aim of the project was to devise a system whereby integrated environmental zones could be defined around extensive industrial complexes, which would have clear spatial consequences.

A 'provisional' version of this system appeared in 1990, which was called the 'Provisional System for Integrated Environmental Zoning (VS-IMZ)'. This methodology made it possible to chart the total (cumulative) environmental impact around complex industrial activities. This cumulative environmental stress could then be translated into an integrated environmental zone that could be implemented in the relevant zoning plan. The consequences for spatial planning would derive directly from the zoning plans and, hence, the system was 'a legally binding or policy-directing instrument, designed to provide a solution for an existing conflict or prevent a new conflict between environmentally intrusive activities and environmentally sensitive functions' (Borst et al. 1995; 21).

Not long after the project started, however, its goals proved to be too ambitious. The environmental/spatial conflicts on which VS-IMZ focused in a series of pilot projects were usually much more complex than anticipated. The instrument proved to be not flexible enough to deal with these complex cases. The Ministry responded by paying much less attention to the system in its second National Environmental Policy Plan (VROM 1993). Eventually, the popularity of the idea of integrated environmental zoning would falter in the face of drastic changes in environmental policy during the 1990s (cf. De Roo 2001; see also Chapter 2). The VS-IMZ project was formally closed in 1997. Despite its failure in practice, however, the VZ-IMZ project and its rationale are still of interest today. Mostly, they can be quite useful in providing a great deal of insight into the environmental situation of an area.

Procedure

The system for integrated environmental zoning differentiates between a *methodological component* and a *procedural component* (VROM 1990). The cumulative environmental stress is determined in the methodological part. Each type of environmental stress is translated into a standardized score, which can be cumulated with the scores for other types of environmental stress. Next, the procedural steps are used to translate the cumulative environmental stress into an administrative environmental zone which also directly shows the planning consequences relating to the zone, which can then be implemented.

The methodological component of the VS-IMZ is divided into two stages. The first stage determines the individual scores for each type of environmental stress. This is referred to as *standardization*. To this end, the individual environmental stressors are arranged into classes (A to E) that express varying degrees of severity. Arranging the individual stressors according to these five classes means that they become comparable. Table 5.3 shows the sectoral classification for existing situations for industrial noise.

Table 5.3 Classification of industrial noise in an existing situation (after VROM 1990)

Category	Class E	Class D	Class C	Class B
Industrial noise in dB(A)	> 65	65–60	60–55	< 55

Similar tables are used for other types of environmental stress. This may mean that a certain area can be characterized as Class C for industrial noise and Class D for toxic substances. A second step is needed to arrive at a cumulative environmental load experienced in a certain area. This second step examines which combinations of environmental impact are present in a certain area. Each area is then positioned in an 'integrated environmental-quality class', based on the accumulation of the various types of environmental stress experienced. It is this integrated class that determines the planning measures. For example, suppose that Class C is counted three times after standardization, Class B once and Class A once. This would result in an integrated classification of III for existing situations. A stricter classification is followed for new situations. This means that there will be different planning consequences for new situations than for existing areas.

The integrated classes for environmental impact, as mentioned above, are linked to planning restrictions. For existing situations, these can vary from no limitations (integrated Class I) to the demolition of houses (integrated Class VI). The first instance refers to a 'white area', the second to a 'black area'. The classes in between are referred to as 'grey areas'. Figure 5.3 illustrates the classification for existing situations.

The methodological part of the approach is complete once the integrated class has been established. In the procedural part, the administrative environmental zone will then be established. This part can also be divided into two steps. The first step leads to a provisional integrated environmental zone. On this basis, administrators are considered to be following a conservative and prudent approach regarding housing plans or the issuing of environmental permits in the area affected by the provisional zone. Direct spatial-planning consequences only follow from the second step, when a definitive integrated environmental zone is established.

Practice

The provisional system outlined above has been implemented in eleven IMZ pilot projects (1989–1997). In most of the pilot projects, the integrated environmental zones overlapped with environmentally sensitive functions, resulting in conflicts between environmentally intrusive and sensitive functions. In some cases, these conflicts were such that no realistic and feasible solutions could be found.

The pilot project of the Arnhem municipality has gone down in history as a classic IMZ example. The industrial estate at Arnhem-Noord has both large and small businesses. Together, these businesses contribute to a situation whereby those living nearby regularly experience nuisance as a result of the environmental

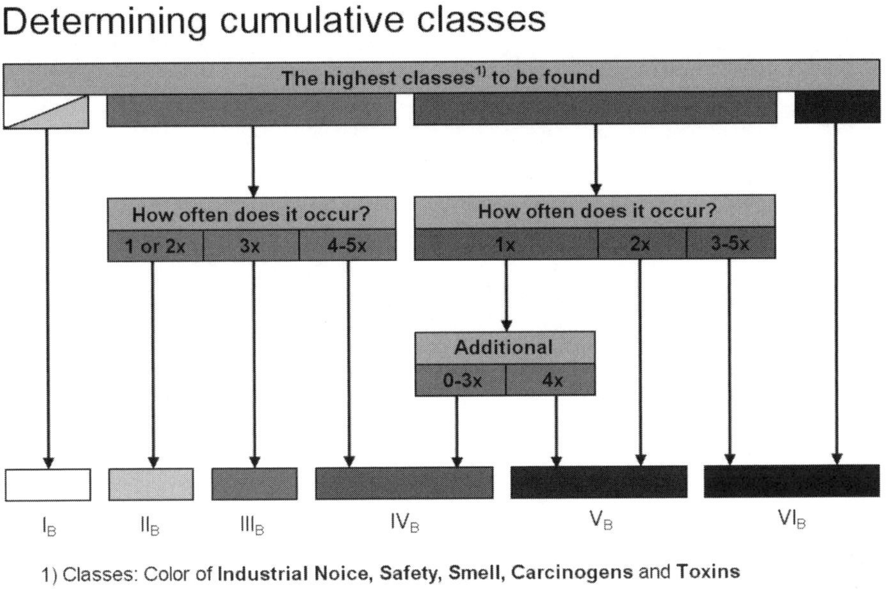

Figure 5.3 Integrated classification for existing situations (b) (after IMZS-Drechtsteden Steering Group 1991)

stress produced. This was reason enough for the province of Gelderland to register the Arnhem-Noord industrial estate as an IMZ pilot project in 1988 (Boei 1993). The assessment revealed that the environmental impact affected a very wide area. In practice, legal limit values for various environmental stressors were exceeded to up to eight kilometres from the estate. Once the integrated environmental zone was determined, it became clear that thousands (!) of houses actually fell into the highest environmental quality class (VI). Characterized as 'black' areas, the (planning) consequences should involve demolition. Obviously, these consequences were not accepted, proving how the outcomes of the pilot project were thus completely unacceptable from an administrative point of view.

The Drechtsteden pilot project also produced serious conflicts (Voorknecht 1993). A series of quite impressive maps (Figures 5.4 and 5.5) emerged from the calculation of the integrated environmental stress. Large parts of the area were located within the Class VI contour lines, both for existing and for new situations. Again, the planning consequences proved unacceptable. The practical applicability of the methodology was again challenged.

The majority of the pilot projects were characterized by complex environmental/planning problems. Although the results for Arnhem and the Drechtsteden were shocking, other pilot projects were also confronted with Class VI contour lines or black areas on the map. The severe planning consequences were mostly considered unacceptable. Hence, final administrative environmental zones, which

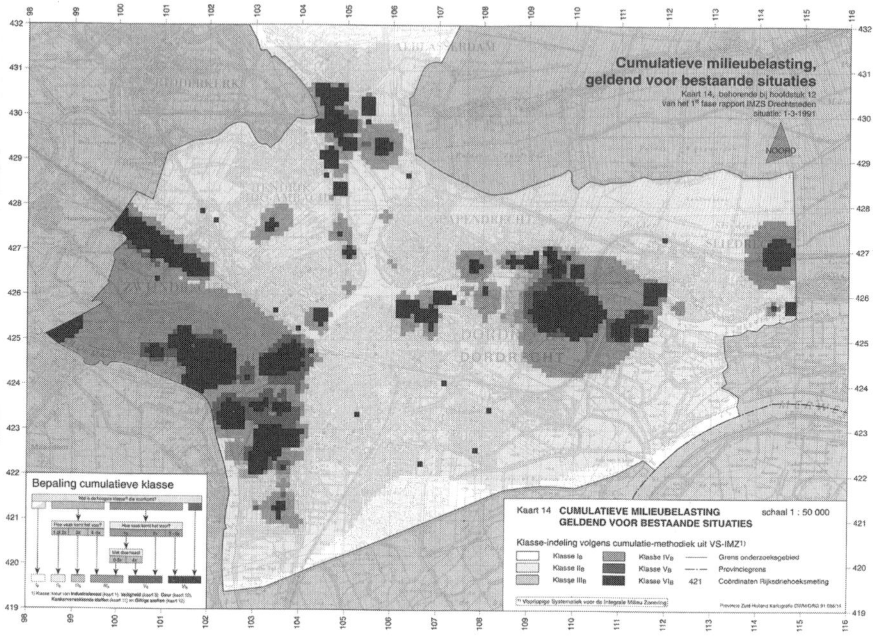

Figure 5.4 Zoning map for the Drechtsteden pilot project, for existing situations (IMZS-Drechtsteden Steering Group 1991)

would make these consequences legally binding, were usually not produced. One project did actually determine and implement an administrative environmental zone. In Hengelo, the integrated environmental zone was processed into various relevant zoning plans. It proved how integrated environmental zoning can indeed contribute to planning policy. Nevertheless, most pilot projects reverted to existing sectoral policies which did not take account of the accumulation of various stressors. Hence, their spatial planning consequences were considerably less far-reaching. Although this might be considered a failure of the VS-IMZ project, it was still successful as a method for making the integrated environmental quality transparent on the basis of actual data.

Discussion

The discussion of the VS-IMZ approach is again divided into three sections. The first section assesses the position of the approach within Dutch environmental policy, based on the framework discussed in Chapter 3. The second section of the discussion involves a more in-depth analysis. The five distinctive characteristics of methods for environmental externalities as discussed in Chapter 3 will all be examined. The last section will concentrate on the applicability of the approach in planning processes.

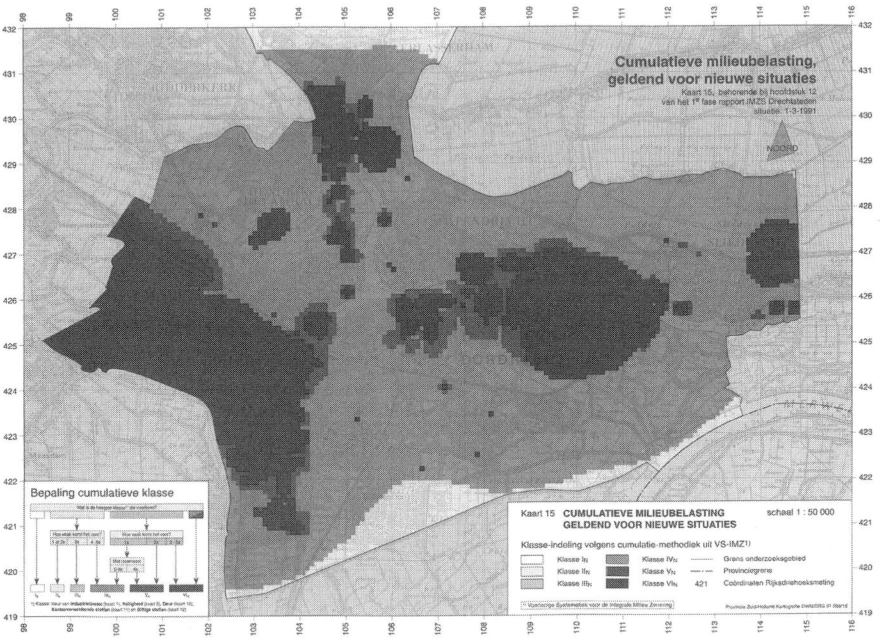

Figure 5.5 Zoning map for the Drechtsteden pilot project, for new situations (IMZS-Drechtsteden Steering Group 1991)

The pilot projects for implementing the provisional system for integrated environmental zoning revealed that stringent application of the planning consequences following from the system is not always feasible from an administrative point of view. The technical and legalistic approach of the methodology, however, is characteristic of the environmental policy of the 1980s and early 1990s. Given this approach, VS-IMZ can be placed into the top left part of the framework (Figure 5.6a), a position that underlines the technical-rational character of the methodology. In addition, this position is strengthened when we remember that the calculation method is based on stringent standards and that the planning consequences linked to the methodology are dictated by central government.

The aim of the 'Accumulation of sources and integrated environmental zoning' project was to arrive at legal regulations for integrated environmental zoning 'to improve the environmental quality around large industrial complexes' (VROM 1989). To this end, the improvement of environmental quality is enforced by the stringent application of environmental standards, on which the entire concept of environmental zoning is based. With regard to the characteristic *intention*, VS-IMZ can be regarded as mandatory and is thus placed in the innermost ring of the model.

One strong aspect of the methodology is the integrated way the environmental stress is expressed, resulting in an integrated environmental zone. The approach only relies on the calculation and accumulation of the types of environmental stressors included in the system. Hence, other aspects that influence the quality of the physical environment are not taken into consideration. Since this undermines its initial integrative purpose, the approach can best be placed in the middle ring when considering the characteristic *integration*.

The intended methodology for integrated environmental zoning is primarily designed for use by local government. Certainly in relation to the methodological component of the approach, there is only limited room for *interaction* with other interested parties. Therefore, the approach is placed in the innermost ring of the model.

The system – and this is where VS-IMZ differs from the VNG approach – makes no allowance for differences in area sensitivity. The starting points are the functions 'living' and 'industry', and there are no happy mediums. Thus *differentiation* based on differences in the sensitivity of an area and its land use to environmental stress is not an option. The approach is therefore also placed in the inner ring of the model for this characteristic.

The last characteristic is *conformance & performance*. The planning consequences of applying the method have been referred to frequently. The integrated environmental zone following from the method is meant to be directly implemented in local environmental policies and spatial plans (usually zoning plans). Given this direct implementation, the approach should be placed into the innermost ring of the model. Figure 5.6b depicts the relationships between the characteristics.

Implementing an integrated environmental zone in spatial zoning plans requires the methodology to be used at an early stage of the planning process. The integrated environmental stress will need to be clear from the start of the preparation of the zoning plan procedures. The applicability of the methodology is wider, however, and extends to the entire decision-making process. VS-IMZ offers an insight into the current quality of an area in terms of environmental health and hygiene. This information can be important at different times during a decision-making process – and elsewhere. Figure 5.6 illustrates the overall picture of VS-IMZ.

VS-IMZ: Conclusions

The overview reveals various strengths and weaknesses. One important strong point of the system is the insight that can be gained into the actual environmental quality of an area. The system is thus suitable for illustrating complex environmental stress in relationship to an area. Another strong point is the way in which the information acquired is presented. Environmental quality is visualized in maps. The discussion of the insight methodologies (Chapter 8) will reveal that this can strengthen the communication (and thereby integration) between environment and spatial planning, and between government, citizens and business.

However, several less strong points have contributed to the eventual failure of the approach as used outside the pilot projects. The planning consequences of the approach are the main contributors to this failure. Several pilot projects showed

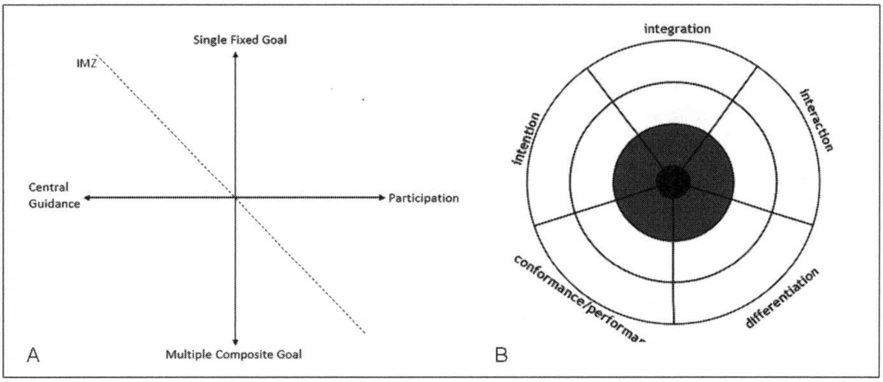

Figure 5.6 Characteristics of the Provisional System for Integrated Environmental Zoning

how the application of the approach would have had dramatic consequences for some of the projects – large areas of the Drechtsteden and much of Arnhem would have vanished from the map. Another less strong point is that the measurements and calculations to be carried out in the methodological component of the system are time-consuming and expensive. Next, the sheer number of integrated classes, particularly for the grey area, is also referred to as a disadvantage. They make the methodology too complicated and unwieldy. Finally, the approach does not allow for differences in the sensitivity of an area and its land use to environmental stress. As this is an important aspect in current environmental policy, the approach would also fall short on this aspect.

As a methodology, VS-IMZ has proved to be unfit to function in the arena of clashing interests. The most important cause is considered to be the stringent planning consequences linked to the procedural part of the methodology. This is one of the reasons why VS-IMZ has been consigned to the waste bin. Nevertheless, despite the strong criticism, the system for integrated environmental zoning should not simply be written off. In particular, the methodological component of VS-IMZ has turned out to provide a reasonably good insight into the environmental quality of an area (Borst et al. 1995). The system helps to make the causes and consequences of various environmental impacts transparent. This transparency is very important for local policymakers when they try to balance various interests. Seen from this perspective, the system still has much to offer. To that end, however, the compulsory character of the methodology should be abandoned. What remains is a type of approach that comes close to the 'information methods', which will be discussed in Chapter 9.

5.3 Environmental Zoning Approaches in Perspective

Environmental zoning is an instrument that has been around for nearly twenty-five years. It is an instrument that – in general – can provide quick and easy information on the current environmental situation in an area. The methodologies discussed in this chapter are known and used nationally. Since the first appearance of the VNG approach, the well-known 'Green Book', it has gone through several editions, and a fourth version is pending. The second approach was developed under the responsibility of the Ministry of VROM. Unlike the popular VNG approach, and despite initially offering more opportunities as well as being a welcome supplement to the VNG approach, VS-IMZ has not been developed further.

When the approaches are compared to each other, various things draw our attention. Although both methodologies are presented as integrated zoning approaches, there is no question of integration in the VNG approach, in which the greatest distance to be maintained defines the environmental zone. The VS-IMZ enables a cumulative environmental stress which forms the basis for the integrated environmental zone. The later version of the VNG approach scores better than VS-IMZ in the possibility of area differentiation. The VNG approach has thus adapted to the requirements of current environmental policy, in which the situation-specific aspects of issues and areas are regarded as important facts.

The VNG approach is still trusted widely. Even today, the approach is being implemented by municipalities. In addition, the Council of State regularly refers to the approach when ruling on environmental cases. It is thus no surprise that the fourth version of the approach is currently in preparation. The story of VS-IMZ is different. But despite its unpopularity, the system should not just be thrown out with the bathwater. The approach may have failed as a framework-setting zoning instrument, but the methodological component, for example, is still suitable for gaining insight into the current environmental situation of an area. This also means that the approach has not been completely forgotten. For example, a recent report by the GGD (municipal health service) for Kennemerland mentions that the approach is usable for gaining *insight* into the dispersal of substances (Oosterlee and Keuken 2004). Therefore, a link to environmental atlases that are being developed is perhaps one of the possibilities.

Despite the promising start and the continuing use of the VNG approach, overall faith in environmental zoning approaches has gradually declined. The belief that the relationship between environmental and planning policy could be regulated by maintaining generically specified distances turned out to be untenable. In the mid-1990s, after the experiences with the IMZ pilot projects, this was expressed at the Nunspeet conference (see Chapter 2, VROM 1995). The situation-specific and area-specific characteristics of issues were then regarded as the keys to the solution of most complex local environmental issues. Although the VNG approach gradually and partially adapted to this, new approaches began to emerge.

Towards the end of the 1990s, the first methods for environmental externalities capable of allowing for these characteristics began to appear – the 'Environmental

Quality Perspectives'. Before we examine these methodologies in detail in Chapter 7, the next chapter will focus on 'checklist' approaches. These are methodologies designed to ensure that environmental interests or health issues are not 'forgotten' in the daily tumult of local planning practice.

Chapter 6
Checklist Methods: Between Quantity and Quality

The role attributed to environmental interests – and hence to nuisance, health issues and risk – in decision-making processes is a much-discussed subject within spatial planning. Paradoxically, the overall feeling is that environmental interests are not emphasized enough. This is strange, since health is considered to be one of the most important aspects of life. Over the years, the national government has tried in various ways to integrate spatial planning and the environment,[1] with varying degrees of success. Integrated Environmental Zoning, for example, might appear at first sight to have been a comprehensive policy instrument. However, a closer study reveals how its top-down enforcement of rigid environmental standards has not produced the desired cohesion between spatial planning and environmental policy. In fact, Integrated Environmental Zoning (IMZ) even put this relationship under greater pressure.

The lack of success with IMZ undoubtedly contributed to the development of new methods for environmental externalities. In the first half of the 1990s, methods evolved which, by means of checklists or control lists, placed the theme 'environment' (and therefore nuisance, risk and health) on the spatial-planning agenda. This chapter focuses on these methods.

The first checklist method we will discuss here is the Environmental Performance System (*Milieuprestatiesysteem*), developed by the Amsterdam municipal authority in 1994 (Section 6.1). The aim of this method was to link these environmental ambitions and qualities to the local zoning plan, so as to 'guarantee' certain environmental standards in spatial developments. Section 6.2 discusses a second method: the *Milieuplaberum* (a planning and decision-making process for spatial measures), which was also developed by the Amsterdam municipal authority. The *Milieuplaberum* indicates at what stage in planning processes it is best to introduce environmental measures. A third checklist method is Health Impact Screening (HIS), developed by the national association of municipal health services (Section 6.3). This instrument made it possible to assess at an early stage the health-related effects of spatial developments. The three methods will be compared in Section 6.4.

1 This was discussed in detail in Chapter 2.

6.1 Amsterdam's Environmental Performance System

Zoning plans are the most important planning instrument that local authorities use with regard to the physical environment. Zoning plans specify how the space in the relevant plan area can be used. The purpose of this type of plan is not only to steer spatial development in the right direction and ensure legal certainty for citizens with regard to spatial interventions, but also to safeguard the quality of the physical environment.[2] However, the Spatial Planning Act (WRO, *Wet op de ruimtelijke ordening*) allowed for the incorporation of environmental aspects to only a quite limited extent.[3] One of the few options available to local authorities for this purpose, was to define and implement spatial zones on the basis of environmental health and hygiene conditions (environmental zoning). However, as we have seen in the previous chapter, this instrument does not always produce the desired result.

At some time during the 1990s, the municipal authority of Amsterdam considered that these options in conventional zoning plans were unsatisfactory, although it considered the role of environmental considerations in spatial-planning processes to be quite crucial. As they found themselves unable to put its ideas into practice due to the lack of 'tailor-made' instruments, it developed its own instrument for this purpose. In 1994, the Policy Document on Spatial Planning and the Environment appeared (BROM; *Beleidsnota Ruimtelijke Ordening en Milieu)* (Gemeente Amsterdam 1994a). In this document, the environment is regarded as one of the key criteria for assessing the quality of the physical environment. In addition, two key observations in the BROM eventually contributed to the development of the Environmental Performance System.

First, the report noted that 'it is increasingly clear (…) that space and the environment have a strong influence on each other' (Gemeente Amsterdam 1994a; 5). Second, it also noted that 'there is a lack of integration-orientated tools' (Gemeente Amsterdam 1994; 5).[4] Consequently, the aim of the BROM was to achieve greater integration of spatial planning and the environment. To that end, the BROM contains an overview of instruments that can contribute to the development of integrated area-specific policy. Among the instruments discussed

2 The zoning plan is the only type of plan in spatial planning that is legally binding in respect of citizens. It not only provides legal certainty for stakeholders in the relevant plan area, but also provides certainties with regard to the quality of the physical environment.

3 Voogd (1999) points to the outline of a zoning plan that in practice appears to be an effective way of, for example, incorporating policy principles for the environment into a zoning plan (see also Lubach 1991, Klaassen 1994).

4 In the zoning plan it drew up for Morra Park in 1993, the Smallingerland municipal authority experimented by adding environmental requirements (Timár 2005). These additions were neither permitted nor prohibited under the Spatial Planning Act of the time. The Amsterdam municipal authority saw this as an opportunity to experiment with a method of its own: the Environmental Performance System.

are the Bubble Method (Chapter 9) and the Environmental Performance System. Furthermore, the Environment Matrix (*Milieumatrix*)[5] – which to a certain extent can be regarded as the forerunner of the *Milieuplaberum* (Section 6.2) – is discussed. Of the methods discussed, the Environmental Performance System seemed to have the greatest potential for contributing to integrated area-specific policy. Eventually, the Dutch Council of State would declare the method and the resulting zoning plan to be invalid. However, this does not lessen the inspirational value of the method's basic premises and logic.

The Environmental Performance System seeks to develop integrated area-specific policies in a very direct and straightforward way. The system is based on the idea that, when zoning plans meet certain environmental targets, they will earn a number of points. Based on the points earned, a project is either considered acceptable and can go ahead, or unacceptable and in need of additional measures to attain the minimum score.[6] The method for calculating points makes the system quite flexible, and it is possible to select a range of measures in order to achieve the minimum score. This means that the system allows for local preferences and can be tailored to the situation at hand.

This section discusses the functioning of the Environmental Performance System on the basis of the *IJ-oevers* zoning plan (Gemeente Amsterdam 1994b). This zoning plan was developed while using the Environmental Performance System, which added direct environmental requirements to the zoning plan. Later we will see how this was eventually dismissed in law by the Council of State.

Procedure

The minimum score is expressed as an 'environmental performance level' (EPL). Each plan has a different EPL that must be specified in advance. This is not based on a particular method, scientific or otherwise. The EPL is chosen arbitrarily.[7] The way in which scores are assigned to the various environmental measures is equally arbitrary. That being said, the EPL and the list of measures (see Table 6.1) form the core of the method.[8]

5 In 1995 the University of Groningen carried out a study of environmental assessment methods (Humblet and De Roo 1995). In this study, the Environment Matrix is compared with the Environmental Performance System. While the latter is designed for zoning plans, the matrix is designed to reinforce the structure plan.

6 The principle of compensation is clearly evident in this method. A target that is outside the capacity of the local authority can be compensated by meeting one or more environmental targets that are easier to achieve.

7 In the notes to the *IJ-oevers* zoning plan, the Amsterdam municipal authority clearly states that 'For the time being, the environmental performance system is a policy instrument, not a scientific method' (Gemeente Amsterdam 1994c; 22).

8 It was intended that the EPL would be legally established in the outline or explanatory notes of the zoning plan. These would also include the list of measures specifically for the planning area in question.

Table 6.1 Environmental measures, targets and scores for the theme 'Mobility' (Gemeente Amsterdam 1994b)

Environmental measure	Target	Points
Theme: Mobility		
Walking distance in metres to nearest public-transport stop	> 300	+1
	> 200	+2
	> 100	+3
Parking standard for residential areas (no. of parking spaces per residence)	< 1.0	+1
	< 0.8	+2
	< 0.6	+3
Parking standard for work locations (no. of parking spaces per 250m^2)	2 (B location)	+1
	1 (A location)	+2
	< 1	+3
Theft-proof bicycle parking within a radius of 100 metres (no. of spaces per residence)	1	+1
	2	+2
	3	+3
Parking spaces in built car parks exchangeable for Call-a-Car facilities (no. of cars)	> 10	+1
	> 25	+2
	> 50	+3
Theme: Water/Green Spaces		
Rainwater drainage (from roofs):	100% into sewer system	-1
	< 100% into sewer system	+1
Take-up of basic water-saving package:	No	-2
	Yes	+2
Construction of planted roofs/roof gardens (excl. terraces)	Yes	+1

Environmental measures that are regarded as relevant to the zoning plan can be selected from the list. In the case of Amsterdam, the list comprised seven themes: compact city, mobility, noise pollution, sustainable construction, energy, water/green areas and waste (Gemeente Amsterdam 1994b). Table 6.1 shows the possible measures that can be taken for the themes 'mobility' and 'water/green areas'. In the column with the scores we see how many points can be lost or earned when a certain measure is applied. When points are lost, the initiator will have to make an extra effort to attain the predefined EPL so that the project can proceed.

The Environmental Performance System partly depends on the principle of compensation. The way in which compensation can be sought is simple. Suppose that, in the draft zoning plan, all rainwater drains into the sewer system. In Table 6.1. we see that this results in the deduction of one point from the total score. In the same plan, an extra point is earned by reducing the walking distance to the closest public-transport stop from 320 metres to an average of 270 metres. The reduced distance to public transport thus compensates for the rainwater drainage.

Inspired by the compensation principle, the Environmental Performance System can contribute by setting both restrictive and progressive conditions with regard to liveability. It is restrictive to suggest that choices which will have a negative influence on the total score should be avoided as far as possible. An example is the draining of rainwater into the sewer system. It is progressive for most other measures. In practice, the initiator is enabled to optimize his choices based on the points that can be earned and the efforts required in order to earn these points. This implies that the zoning plan can be more easily enhanced with liveability and sustainability aspects.

Practice

As said above, the Environmental Performance System was used for the first time in the *IJ-oevers* project on the banks of the river IJ in Amsterdam. This project involved the restructuring of the southern banks of the IJ in order to connect them to the centre of Amsterdam. The terms of reference for the project stated that 'Metropolitan developments such as the *IJ-oevers* project require an ecological approach' (Gemeente Amsterdam 1991; 95). On the basis of this premise, the integration of environmental and spatial considerations was considered an important foundation for the restructuring project.

This was confirmed when the *IJ-oevers* zoning plan was published in 1994 (Gemeente Amsterdam 1994). The environmental component is clearly present in the plan's outline, explanatory notes and appendices. The integration of environmental considerations in the zoning plan is a direct result of the link to the Environmental Performance System. The minimum ELP score for the *IJ-oevers* zoning plan was set at 20. In the most ideal situation (maximum environmental measures), it would be possible to achieve a total score of 47. In using the method, the municipality committed itself to meeting the minimum EPL score. Seen from this perspective, the Environmental Performance System guarantees a certain level of integration of the environment in spatial plans.

However, the use of the system in the *IJ-oevers* project led to a confrontation with the Council of State. The Administrative Jurisdiction Division of the Council of State declared the linking of the Environmental Performance System to zoning plans to be unlawful. This meant that it was not possible to use the method within the framework of the classic Spatial Planning Act.[9]

Discussion

The discussion of the Environmental Performance System has three parts. The first section assesses the position of the approach within Dutch environmental policy, based on the framework discussed in Chapter 3. The second section explores the

9 There have since been calls to include environmental-quality requirements in zoning plans (see for example VROM 2006). This is discussed later in the book.

Environmental Performance System in greater depth, while addressing each of the five characteristics of methods for environmental externalities. The third section focuses on the method's scope of application in planning processes.

We assume that the development of environmental policy can largely be illustrated as a shift along the diagonal of the framework (see Section 3.1). A position in the upper left-hand quadrant is associated with the wish to achieve a maximum policy result. An example is the use of environmental standards that are imposed top-down and must be complied with in all cases. A position in the lower right quadrant relates to the wish to follow an optimum policy process, thereby reflecting decentralized environmental policy. The Environmental Performance System can be positioned in the central section of this spectrum (Figure 6.1a). On the one hand there is a hard, almost standards-based goal: the EPL (environmental performance level). On the other hand, compensation and local 'customization' are possible because a selection can be made from a range of measures in order to meet the EPL. The flexibility to select various combinations of measures contributes to finding a more optimum set of measures to enhance the quality of the environment.

When considering from the perspective of each of the five characteristics of methods for environmental externalities, a varied picture emerges. The characteristic *intention* is mainly expressed in the EPL (environmental performance level), which is defined in advance and is a minimum requirement for project approval. From this point of view, the method can be regarded as mandatory and is therefore placed in the innermost ring of the model.

A second characteristic is *integration*. The method is designed to integrate environmental requirements in spatial plans. There is free choice with regard to which requirements to include. In the case of Amsterdam, for example, measures relating to external safety and odour nuisance were not included in the measures from which a selection can be made.[10] However, this was a decision made by the Amsterdam municipal authority. The method does not place a limit on the number of environmental requirements to be included. The Environmental Performance System therefore belongs in the outer ring of the model.

A third characteristic is *interaction*. The method is intended to be part of a larger project, usually the development of a zoning plan. Interaction with stakeholders takes place through statutory consultation procedures. There is, therefore, definitely scope for interaction – albeit to a limited extent. This means that the method can be positioned in the middle ring of the model.

The fourth characteristic is *differentiation*. Indirectly, area differentiation is possible, for example by taking account of the characteristics of an area when selecting measures. Direct differentiation (e.g. by defining spatial zones) is not

10 This is even more surprising when we consider that the method was first put into practice in an area (the banks of the river IJ in Amsterdam) where noise pollution and odour nuisance were certainly relevant environmental themes (compare De Roo 2001).

among the options. In terms of the characteristic *differentiation*, the method can be placed in the middle ring.

The final characteristic is *conformance & performance*. The environmental measures selected by means of the Environmental Performance System must be legally established in the zoning plan. This means that there is a direct outcome, so the method is placed in the innermost ring of the model. Figure 6.1b shows the total picture of the five characteristics for the Environmental Performance System.

The third part of this discussion focuses on the applicability of the approach. There is no statutory requirement to apply the Environmental Performance System to certain spatial projects. It is used on a voluntary basis. However, once the decision has been made to use the system, certain obligations must be fulfilled. For example, the method must be initiated at an early stage in the planning process. Only then, it is assumed, can environmental requirements be effectively incorporated in the spatial-planning process. The Environmental Performance System is also important in the later phases of the planning process, for example, in order to evaluate choices and hence monitor the input of environmental aspects in the planning process.

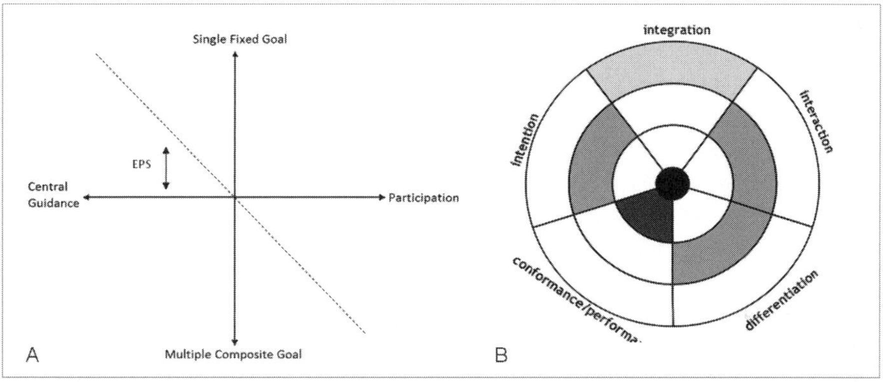

Figure 6.1 Characteristics of the Environmental Performance System

The Environmental Performance System: Conclusions

The discussion reveals various strengths and weaknesses. One of the strengths of the Environmental Performance System is that it is a simple way to monitor the integration of environmental considerations in spatial planning processes. Another strength is that, depending on the conditions, the list of measures can be tailored to individual situations. The approach is therefore suitable for different urban areas, from inner cities to industrial areas. A third strength is the contribution made to restrictive and progressive measures for liveability. It is restrictive in the sense that measures that have a negative impact on liveability will be less readily chosen (deduction of points). On the other hand, choices that have a positive impact on liveability will also have a positive effect on the score. In this way, the initiator is encouraged to make the best possible choices. Finally, there is the possibility of using the method to develop sustainable spatial plans. As we have said, the measures can be tailored to individual situations and it is therefore possible to introduce sustainability aspects into spatial plans.

However, the Environmental Performance System also has a number of weaknesses. The first of these relates to the substance of the method. The minimum EPL to be attained is set in an arbitrary way. This also applies to the scores to be achieved for environmental measures. This problem might be addressed if another form of substantiation is developed or if the measures and related minimum scores are defined in advance. Another weakness relates to the level of support for the method in the administrative sphere. Within the framework of the classic Spatial Planning Act it was not possible to incorporate environmental aspects in spatial plans. On that basis, the Council of State declared the link between the Environmental Performance System and zoning plans to be unlawful.

In abstraction, the Environmental Performance System is a simple and therefore cost-effective method for adding environmental aspects to projects in the sphere of spatial planning. The method does not necessarily contribute to solving concrete issues, but sets out an ambition for new areas and areas to be restructured. It provides a range of measures and options that can help to achieve a certain level of quality in the physical environment. The method would therefore not be out of place in current environmental policy, in which achieving a balance between diverse interests is an important factor. However, the ruling by the Council of State in 1999 pushed the Environmental Performance System into the background.

Yet there are still opportunities for the Environmental Performance System. In the Future Agenda on the Environment (*Toekomstagenda Milieu*), the Ministry of VROM again underlines the importance of achieving coherence between spatial-planning policy and environmental policy. To that end, the Agenda proposes that local authorities should be given additional opportunities for including environmental-quality requirements in zoning plans (VROM 2006; 37). However, this will require quite a few changes to the legal framework. A first step in that direction has already been taken with the implementation of the new Spatial Planning Act (WRO, *Wet op de Ruimtelijke Ordening*) in 2008.[11] The explanatory memorandum of the new WRO expressly states that the bill allows for the interests of other policy areas to be considered when zoning plans are drawn up (TK 2002; 8). However, while developments are thus increasing the possible use of the Environmental Performance System, doubts remain. New developments in Dutch spatial legislation have to be approved by the Council of State. In line with what

11 The current WRO dates from 1965 and has been amended many times. In May 2003, a new WRO bill was submitted to the Lower House of Parliament and was approved on 23 February 2006. The bill was approved by the Senate on 17 October 2006, leaving the way clear for implementation.

happened in 1999, the Council of State has expressed, in advice to Dutch policy–makers, that it has reservations about this development.[12]

However, with the new WRO and the scope it created for environmental requirements in zoning plans, the government has sent out a clear message:[13] the coordination and coherence of spatial-planning policy and environmental policy is essential for the quality of the physical environment. This cannot be achieved simply by imposing restrictions, as in traditional permission-based planning. It requires more than this. In the Netherlands, the concept of '*ontwikkelingsplanning*' (development planning) is used to signify this altered attitude towards the role of policies and regulations. Instead of allocating, zoning and protecting through regulations, the idea is now to use policies to encourage and facilitate cooperation and design (Dammers et al. 2004). This shift requires a new approach, and the Council of State does not appear to be ready to make such a change.[14] Nevertheless, the scope for environmental-quality requirements is increasing.[15] Considering these developments, there are also new opportunities for the Environmental Performance System.[16]

6.2 *Milieuplaberum*

In the previous section, the Environmental Performance System was described as a checklist for giving environmental considerations a place in zoning plans. This experiment failed at the time, partly as a result of legal restrictions. In 1997, the *Mileuplaberum* was introduced and to a certain extent can be regarded as the successor to the Environmental Performance System. The *Mileuplaberum* also comprises a list of environmental measures which – and this is one of the differences – can be introduced in phases during the course of the spatial planning

---

12 The Council of State was of the opinion that coordination with other areas would become more complicated and eventually have a negative impact on implementation (Raad van Staat 2002).

13 Apart from the WRO, the Environmental Management Act (*Wet Milieubeheer*) also had to be amended to allow for the inclusion of environmental-quality requirements in zoning plans.

14 This is all the more surprising given the fact that the Council of State did accept the VNG environmental-zoning method. In this way too, environmental aspects are introduced into zoning plans (see Section 5.1).

15 The Minister for VROM responded to the Council of State's advice in a follow-up report (VROM 2002c), stating that the explanatory memorandum has been amended so that the scope is clearly delineated. Time will tell whether this is the case in practice. If the delineation is not clear enough, there is a risk that citizens will lodge objections to the zoning plan if the environmental-quality requirements are not sufficiently transparent.

16 Before we reach this stage, the method's underlying reasoning requires some attention. In particular, the way in which the scores are determined for the measures is dubious.

process. A second difference is that, unlike the Environmental Performance System, the *Mileuplaberum* has been successful in policy practice.

The *Milieuplaberum* can be regarded as a differentiation of the *Plaberum*: the *Plan- en Besluitvormingsproces Ruimtelijke Maatregelen* (Planning and Decision-Making Process for Spatial Measures) dating from 1984 and updated in 2003.[17] In the municipality of Amsterdam, this approach is widely known and used in spatial planning (Gemeente Amsterdam 1999). The *Milieuplaberum* serves as a guideline for introducing environmental aspects into the *Plaberum*. In addition, the method is intended as a checklist designed to remove doubts during planning processes as to the question 'Have we thought of everything?' (Gemeente Amsterdam 1997; 9).

The history of the *Milieuplaberum* goes back to 1993, the year in which the first steps were taken towards the external integration of various policy sectors with each other in Amsterdam. One of these steps was the *Checklist for the integration of environmental aspects in Amsterdam spatial planning*. Following several updates of the checklist, the *Milieuplaberum* was introduced in 1997. It is a logical follow-up to the 1993 checklist. The *Milieuplaberum* was presented at the same time as a large number of other instruments for local environmental policy.[18]

Procedure

The name *Milieuplaberum* suggests that the instrument has many aspects in common with the *Plaberum*, and that is indeed the case. The Plaberum is, in fact, a common process structure that is used in spatial-planning processes by the municipality of Amsterdam. It comprises seven phases (0 to 6), structured in order to ensure that various *spatial* measures are implemented at the most appropriate time. The *Milieuplaberum* – a differentiation of the *Plaberum* – is designed to introduce environmental measures in the most appropriate *Plaberum* phase (Gemeente Amsterdam 1997).

Two components can be distinguished in the *Milieuplaberum*. In the first place, there is the process structure, which is set out in seven phases: the initiative (Phase 0), the preliminary study (Phase 1), the Terms of Reference (Phase 2), the Schedule of Requirements (Phase 3), the Urban Plan (Phase 4), realization

17 In the new version of the *Plaberum*, the number of phases has been reduced from seven to four (see for example Gemeente Amsterdam 2006). The discussion in this section relates to the seven-phase version.

18 The Policy Document of Spatial Planning and the Environment (BROM, *Beleidsnota Ruimtelijke Ordening en Milieu*) was published in 1994. The document presented many instruments for achieving coherence between spatial planning and the environment (Gemeente Amsterdam). Four years later, these instruments were evaluated. This led to the publication of the Guide for Spatial Planning and the Environment (GROM, *Gids voor Ruimtelijke Ordening en Milieu)*, which was oriented to the more effective use of the instruments. The GROM included the *Milieuplaberum* and the Environmental Performance System.

(Phase 5) and, finally, the management phase (Phase 6). The second component of the *Milieuplaberum* is a thematic list of environmental measures that can be selected for implementation in the various phases. Table 6.2 shows part of the list of measures from the *Milieuplaberum*.

Table 6.2 Environmental aspects – companies and environment (Gemeente Amsterdam 1997)

Environmental aspects	P0	P1	P2	P3	P4	P5	P6
Promote water and rail transport	i	f	R				
Policy implementation (ABC locations)	i	v	R				
Inventory of zones and licences – company list		i/v		R			
Corporate environmental management					i	v	R
Transport management				i		v	R
Management plan for industrial estates					i	v	R

An 'i' in the table refers to the phase in which the intention regarding the aspect in question is to be included in the plan. If the local authority intends to promote water and rail transport, we can see from Table 6.2 that this should be dealt with in the first phase (Phase 0). The 'v' indicates the phase in which a variant or design must be approved/adopted. In the case of promoting water and rail transport, the decision should be taken in the second phase (Phase 1). In the third phase (Phase 2), the chosen variant is introduced into the Terms of Reference. This is shown by an 'R' in Table 6.2. In this way, the *Milieuplaberum* takes the involved parties through the planning process step by step, indicating which activities are required in order to incorporate a particular environmental measure in the spatial plan and to ensure its implementation in practice. Furthermore, the list of environmental measures can also indicate which local-authority department should take the initiative for the measure in question.

Practice

All major spatial projects in the municipality of Amsterdam are prepared and implemented using the *Plaberum*. The stages at which decisions must be taken are more or less fixed, thus providing the actors with a structure to work with. Environmental measures are incorporated by means of the *Milieuplaberum*. Both the *Plaberum* and the *Milieuplaberum* have been widely used. The experience gained through their use is positive, while significant improvements have been achieved over the years in terms of coordinating spatial planning and environmental interests at the local level.

Discussion

The discussion of the *Milieuplaberum* has three parts. The first section assesses the position of the approach within Dutch environmental policy, based on the framework discussed in Chapter 3. The second part of the discussion involves a more in-depth analysis in which the five characteristics of methods for environmental externalities will be examined. The third section focuses on the method's scope of application in planning processes.

We assume that the development of environmental policy can largely be illustrated as a shift along the diagonal of the framework (see Section 3.1). It is not easy to establish the position of the *Milieuplaberum* within this framework. The method consists of a system of various agreements and procedures related to various phases in decision-making processes. Each of these agreements and procedures involves different decisions and topics. Hence, they differ strongly. This means that it is essentially impossible to make an unequivocal statement on the position of the *Milieuplaberum* in the spectrum. To address this uneasy situation, it is at least possible to compare the *Milieuplaberum* with, for example, the Environmental Performance System or Health Impact Screening (Section 6.3). On that basis, it becomes clear how the *Milieuplaberum* is less standard-oriented, but sufficiently so to place it in the second quadrant of the framework (Figure 6.2a). This position is less extreme than that of the Environmental Performance System.

There are various features which suggest that the method is rather hierarchical. However, it is not so 'hard' that the intention can be described as mandatory. In the GROM, the municipal authority concludes that the *Plaberum* is seldom followed to the letter. Rather, it is supposed to be a guideline (Gemeente Amsterdam 1997). This means that, in terms of the characteristic *intention*, the method can be placed in the middle ring of the model.

A second characteristic is *integration*. The *Milieuplaberum* is also referred to as a 'catalogue of measures', referring to the comprehensiveness of the environmental measures it includes. The method is designed to facilitate the incorporation of these measures in spatial-planning policy. In terms of the characteristic *integration*, the method can thus be positioned in the outer ring of the model.

In the *Milieuplaberum*, there is limited *interaction* between local-authority departments. In the case of Amsterdam, this relates to the council offices of the city boroughs and the city's central municipal environmental departments. Also, interaction with other groups is fairly limited, although essentially possible. In terms of the characteristic interaction, the method can thus be placed in the middle ring.

The *Plaberum* is described as a model for guiding planning processes. Agreements, fixed decision-making stages and environmental measures are intended to result in the final outcome: a solution to the spatial issue. Differentiation on the basis of differences in area sensitivity is not part of this process. That is why the characteristic *differentiation* has not been filled in.

The final characteristic, *conformance & performance* is a relevant characteristic for the *Milieuplaberum*. The aim of the instrument is to introduce environmental measures into spatial-planning processes at the appropriate stage. This means that there is a direct result. In terms of the characteristic *conformance & performance*, the method can be positioned in the innermost ring of the model. Figure 6.2b depicts the relationships between the five characteristics.

In the third part of the discussion, we examine the possibilities for applying the methods discussed above in the various phases of the decision-making process. Given that the *Milieuplaberum* is a model for decision-making processes, we can say that it is appropriate for all phases. This is also evident from the discussion of the procedure involved. The method offers 'handles' throughout the procedure, from initiation to follow-up. Figure 6.2 gives a total picture of the aspects covered in this discussion.

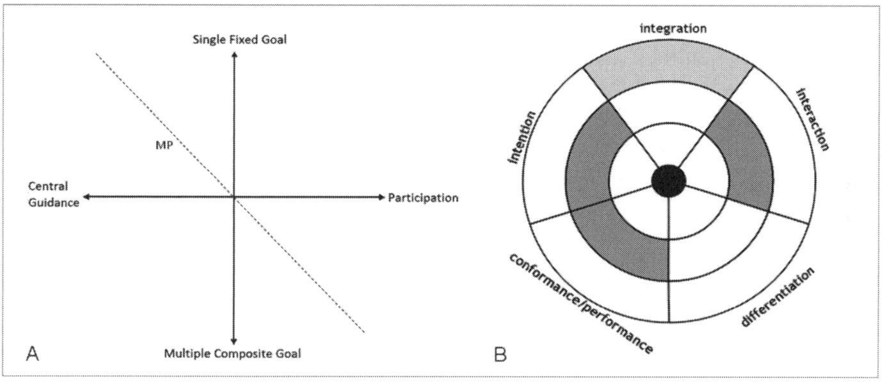

Figure 6.2 Characteristics of the *Milieuplaberum*

Milieuplaberum: Conclusions

The discussion of the *Milieuplaberum* reveals various strengths and weaknesses. One of its main strengths relates to objectives. The *Plaberum* was designed to introduce certain measures into spatial-planning processes at the appropriate stage. The 'environmental variant' of the *Plaberum* does the same thing, but from the perspective of environmental interests. Related to this is a second strength: the *Milieuplaberum* guarantees that the environment is not forgotten, but is brought to the attention of those involved at an early stage.[19] A further strength is the comprehensiveness of the list of environmental measures. The list is compiled in such a way that environmental considerations are highlighted from various perspectives during the various *Plaberum* phases. It is also possible to add to the

19 In the Future Agenda for the Environment, the Ministry of VROM stated that an approach that allows for an integral balance to be achieved between environmental aspects and spatial interests has considerable potential in terms of achieving optimum liveability (2006; 36). The use of the planning method over several years has shown that it achieves results.

list of measures. The clarity in terms of the decision-making stages ('i', 'v' and 'R', see Table 6.2) is also an advantage. These indications enable clear points of action to be drawn up in the planning process, thereby providing guidance for the user. The method was developed by the Amsterdam municipal authority and is hence tailored to its administrative structure. The list of environmental measures also specifies which local-authority department or body must take the initiative for the measure in question. The allocation of responsibility is therefore direct and transparent.

However, the *Milieuplaberum* also has a number of weaknesses. The first of these relates to the list of environmental measures. The list is merely indicative. The officials who implement the method can easily decide to bypass certain measures. They may consider doing so if, for example, the costs relating to a measure are too high or if the timescale involved is too short. As a result, there is a risk that relevant environmental aspects will not be incorporated in the planning process, and environmental interests will not be given sufficient consideration. A second weakness is the fact that the method is geared entirely to the policy structure of the Amsterdam municipal authority. Although this is a drawback, it can be easily resolved by adapting the method for general use.

The *Plaberum* is now more than twenty years old. Since its introduction, the Amsterdam municipal authority has used it to structure its spatial-planning processes. In the mid-1990s, the *Plaberum* was supplemented with a list of environmental measures, the *Milieuplaberum*. This instrument has also proved its worth over the years. Like the Environmental Performance System, the *Milieuplaberum* is an excellent means of introducing environmental considerations into spatial-planning processes at an early stage. And, in contrast to the Environmental Performance System, the *Milieuplaberum* has proved successful as a checklist in practice.

One of the method's most interesting features is its flexibility. Environmental measures can be added or omitted as the situation requires, and the phasing in the *Plaberum* is not written in stone. In the GROM, the Amsterdam municipal authority states that the *Plaberum* is seldom followed to the letter and that 'its application is and will remain a matter of customization' (1999; 7).

These features make the *Milieuplaberum* an interesting method, also in current environmental policy. It incorporates safeguards into the planning process in such a way that environmental measures can be introduced at specified stages. This is a stimulus for the coordination of environment and spatial planning at local level.

6.3 Health Impact Screening (HIS)

The health consequences of a spatial plan or project are not always evident in the early stages of decision-making. As of 2004, large-scale spatial projects in the EU require a Strategic Environmental Assessment (SEA) to be carried out,[20] which considers the environmental consequences of the intentions expressed in plans. An SEA is not compulsory for smaller – local – spatial projects. Without such an assessment, it is difficult to estimate the nuisance from certain sources. This can mean that the level of nuisance caused turns out to be much higher than estimated in an early phase of the planning process, undermining the feasibility of an almost finalized plan (GGD Nederland 2004).

This prompted the Ministry of Public Health, Welfare & Sport (VWS), in cooperation with the national association of municipal health services, to introduce City & Environment Health-Impact Screening (GES, *Gezondheidseffect Screening Stad & Milieu*).[21] The idea behind this instrument is to estimate in advance the health impact of policy intentions so as to reduce the likelihood of frustration and unnecessary costs at a later stage.[22]

The development of Health Impact Screening as a policy instrument has followed an unusual path. It was initially developed as a *qualitative* method. The qualitative HIS method has become a checklist for identifying possible health-impact problems in spatial plans. After the list had been used in practice, the need emerged for a quantitative and uniform method for assessing effects (Fast 2002). The Ministries of VWS and VROM consequently took the initiative to develop *quantitative* HIS. This method, known as City & Environment HIS (*GES Stad&Milieu*) was presented in 2000. In 2002 the HIS method was updated for the first time, largely as a result of certain failings that emerged in practice.[23] HIS City & Environment was reviewed for the second time in 2004 (Fast et al. 2004).

In this book, HIS is categorized as a checklist method, primarily on the basis of its qualitative aspects. The quantitative component of the method shows striking similarities to VS-IMZ (integral environmental zoning) and the 'Environmental Atlas' concept.[24]

---

20 The European SEA Directive (2001/42/EC) was adopted in July 2001. The Dutch SEA system is a direct result of this directive.

21 During the development phase of this instrument, the aim was to work towards a method that would be applicable in the context of the City & Environment experiments. This decision was made on the basis of the reference to HIS in the City and Environment Experiment Act (*Experimentenwet Stad&Milieu*) of the time (Van der Loo & Van Bruggen 1999). Health Impact Screening is now compulsory in Step 3 of the City & Environment approach.

22 Health Impact Screening is carried out on a voluntary basis, as with many other methods for environmental externalities. However, once the decision has been made to carry out screening, certain obligations must be met. This means that it will not always be widely supported within the administrative sphere. This will be discussed further in Chapter 12.

23 These failings related mainly to the starting points for the technical calculations. This aspect will not be discussed further.

24 Chapter 8 argues that the maps produced through HIS can easily be included in an Environmental Atlas (for a discussion of this, see Visser and Zuidema 2007).

Procedure

The HIS procedure is described in a handbook (*Handboek Gezondheidseffectscreening*, Fast et al. 2004). The handbook is primarily intended for use by municipal health services when carrying out a screening for a local authority. Quantitative (numerical) HIS does not begin until the qualitative HIS process has been completed. The full screening process consists of eleven steps.

Health Impact Screening begins with the question of whether screening can help to identify whether, and if so to what extent, the intended measures in a plan will have an impact on public health. In response, the parties involved assess whether health is a relevant aspect in the project in question, whether HIS can be carried out promptly and whether the municipal health service has sufficient resources for this. If the first step is carried out satisfactorily (i.e. HIS is deemed to be of added value for the project), the actual screening process can begin.

In the next steps, checklists are used to identify the sources of nuisance and pollution and to identify the environmental aspects that are relevant to the situation. The different sources of environmental stress that are identified in the HIS process are as follows: railway lines, busy roads, airports, high-voltage power lines, polluted land, and companies that cause external safety risks, odour or air pollution.

Identifying the relevant sources and related environmental aspects constitutes the qualitative component of the Health Impact Screening process. After this is done, the quantitative part of the process can begin. This involves establishing the HIS scores, which form the information that serves as the basis for graphs and contour maps. In doing this, the measurements of the various types of environmental stress are standardized. In this respect, the method is similar to VS-IMZ (Section 5.2). All in all, following the full HIS procedure is quite a technical exercise, particularly in terms of the calculations required to translate measurements into integral HIS scores (Table 6.3).

Table 6.3 Integral HIS scores and related environmental/health qualities

HIS score	Environment/health quality	Colour code
0	Very good	Green
1	Good	
2	Reasonable	Yellow
3	Fairly poor	
4	Poor	Orange
5	Very poor	
6	Unsatisfactory	Red
7	More than unsatisfactory	
8	Very unsatisfactory	

Health Impact Screening produces two types of outcome. First, the report contains graphs and charts of all sources of environmental stress and the associated types of environmental stress caused, along with the related HIS and residence scores. The different HIS scores are summarized in graphs and charts. It is therefore easy to identify the various environmental aspects that are problematic (Figure 6.3). Second, the report contains spatial maps of the plan area, showing the HIS contours. These contours are a direct graphic representation of the integral scores obtained from the screening process and indicate, for example, how many people in the area are exposed to the nuisance or pollution (Figure 6.4). This in turn indicates how sensitive the area is in relation to other areas. Because colour shading is used in the maps, they help to visualize the situation, while, generally speaking, maps are also more readily understood than graphs and charts.

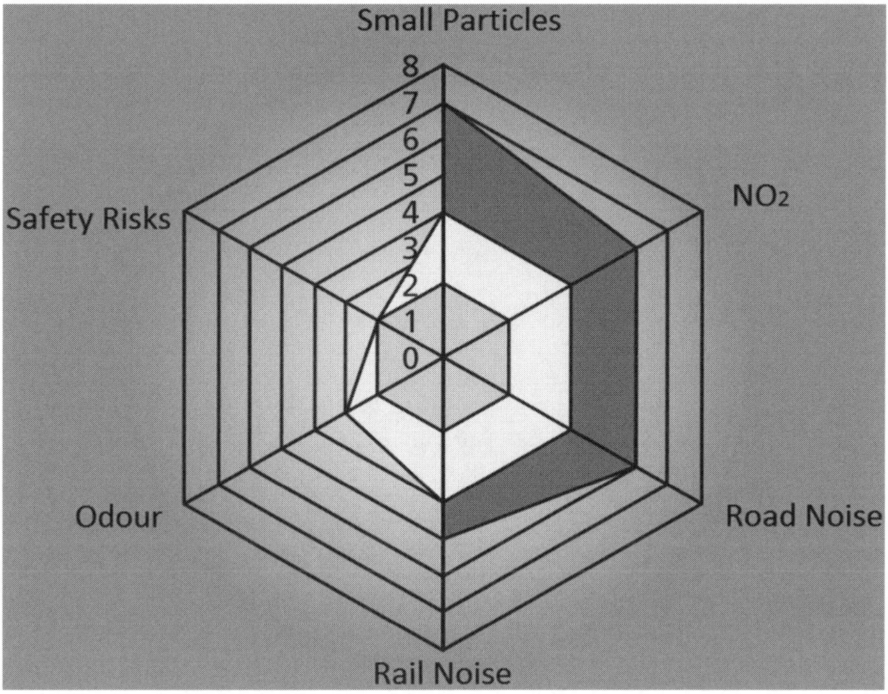

Figure 6.3 HIS scores in a radar diagram (GGZ South Holland)

Practice

Evaluations have shown that, in 2002, approximately 22 percent of Dutch municipal health services had carried out Health Impact Screening or were doing so at the time of the evaluation (Fast 2002). The evaluation report includes an overview of municipal authorities that have carried out HIS for their plans. One such authority is Dordrecht. In 2001, the municipal authority took the decision to develop a 'health park'. The intention was to combine this with the extension of the Albert Schweitzer hospital. Apart from the hospital, the site would accommodate the

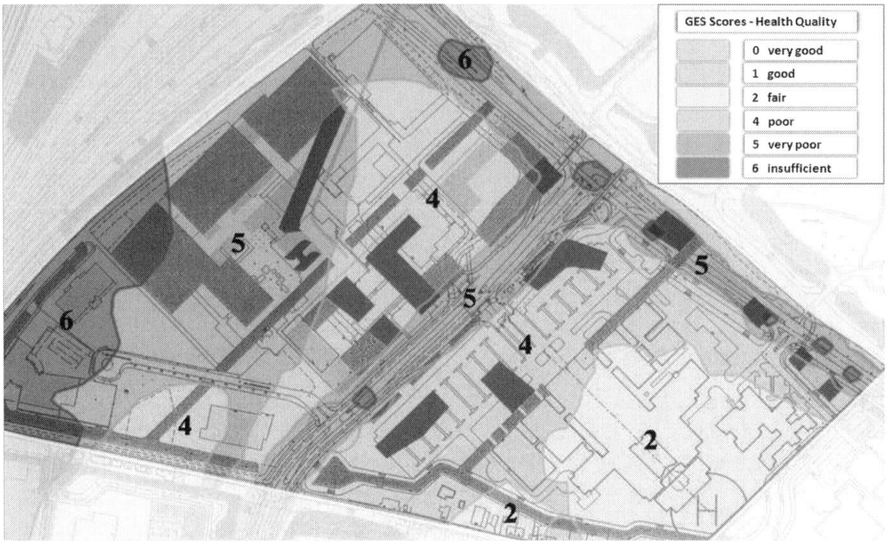

Figure 6.4 HIS scores for Dordwijk Health Park shown as contours (GGZ South Holland)

municipal health service, a general practitioners' centre, home-care service and a sports complex. These are potentially functions that are sensitive to environmental stress. Initially, it appeared that the location of the park would be a problem: a busy traffic route, a railway line and various other local roads were having a serious impact on environmental quality. The local authority therefore requested the municipal health service to carry out HIS (Akkersdijk 2003). The relevant environmental aspects were identified and assessed by means of the screening.

In the case of Dordrecht, the HIS results showed that there were high concentrations of fine particles (PM10) and nitric oxides along the N3 (a busy traffic route). The HIS report expressly advised against locating environmentally sensitive functions close to the road (Akkersdijk 2003). HIS showed that identifying environmental effects at an early stage can be useful for the continuity of the planning process. The Dordrecht municipal authority incorporated a number of recommendations from the screening in its plans (Akkersdijk 2006).

Discussion

The discussion of HIS has three parts. The first section assesses the position of the approach within Dutch environmental policy, based on the framework discussed in Chapter 3. The second part of the discussion involves a more in-depth analysis, again based on the five characteristics of methods for environmental externalities as discussed in Chapter 3. The third section focuses on the method's scope of application in planning processes.

When assessing the position of HIS in relation to Dutch environmental policy, its standards-based character is the first unmistakeable feature to emerge. The contours and areas derive directly from the HIS scores, which are ultimately based on environmental standards. A second striking characteristic can be seen in the pattern of interaction, i.e. there is virtually no interaction. It is the municipal health service that carries out the screening, and the municipal authority – as the commissioning party – makes use of the results from the HIS process. The method thus misses out on the opportunity to involve the public (perceptions) and publish the maps. On the basis of these observations, the method can be placed in the second quadrant of the framework (top left, Figure 6.5a). The meaning of this position has already been discussed for environmental zoning methods.

The picture is further elaborated when it is analysed according to the five characteristics. First, the purpose of HIS is to identify at an early stage the health-related consequences of spatial plans. Ultimately, the goal is for the municipal authority involved to take on the recommendations, but this is not compulsory.[25] Consequently, it is not a mandatory approach. Rather, the commissioning party can itself determine which result to use. Hence, in terms of the characteristic *intention*, the method can be positioned in the middle ring of the model.

Integration refers to the extent to which the method considers environmental aspects in relation to each other. HIS does this to a limited extent. When screening is carried out by a municipal health service, a limited number of environmental health and hygiene aspects are looked at. These are mainly the aspects that have a direct bearing on public health. The screening does not include environmental aspects that are relevant for the quality of the physical environment but do not directly influence human health. This puts the method in the middle ring for the characteristic *integration*.

As remarked earlier, HIS is carried out by a municipal health service without the involvement of potential stakeholders. This makes it easy to indicate where the method should be placed in terms of the characteristic *interaction*: in the innermost ring.

The HIS maps ultimately produced show how many people fall within the contours. They can also indicate whether vulnerable groups are present. This means that, when the results are interpreted, the sensitivity of the area can be taken into account. In terms of the characteristic *differentiation*, the method can be positioned in the outer ring of the model.

Finally, when looking at *conformance & performance*, the information obtained from the HIS process is designed to be used for assessing plan intentions. The information can be made available promptly due to the short throughput time for full screening (ten to fifteen days on average). Obviously, the municipal health service will be keen to see the recommendations adopted. After all, public health is at stake. This puts pressure on a direct implementation of the results. However,

25 HIS is compulsory only in the rare event of a Step 3 procedure according to the City & Environment legislation (only used three times in the Netherlands).

HIS is commissioned on a voluntary basis, and the municipal authority is not required to follow up the recommendations. If other – weightier – interests are at stake, the municipal authority can decide to disregard the recommendations. In terms of the characteristic *conformance and performance*, the method can thus be positioned in the middle ring of the model. Figure 6.5b shows the relationships between the characteristics of HIS.

The scope of application of Health Impact Screening extends mainly over the initial phases of a planning process. Amending plan intentions is relatively straightforward and relatively cheap during these early stages. Consequently, the method must be used during the phases prior to actual decision-making. It can, however, also be implemented in ways other than those envisaged by its originators. Earlier in this section, we mentioned the similarity with 'Environmental Atlases'.[26] The informative maps produced in the HIS process can also be used in later stages of the planning process. An evaluation could be carried out in order to compare new measurements and calculations with previous information. This will be discussed further in subsequent chapters.

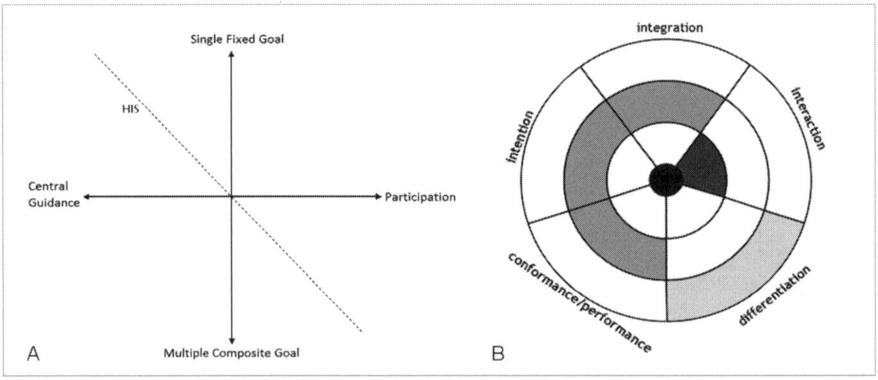

Figure 6.5 Characteristics of Health Impact Screening

Health Impact Screening: Conclusions

The discussion reveals various strengths and weaknesses in the HIS method. One of its strengths is the knowledge and insight it provides into factors that influence human health. This aspect is often underemphasized in Dutch planning practice. Another strength is the speed of carrying out HIS. Depending on the scale of the project and the relevant environmental aspects, full screening takes approximately fifteen to twenty working days. The required information can therefore be available

26 It is therefore interesting to link these instruments. For example, the maps produced in the HES process could be included as 'health maps' in an Environmental Atlas. This will be discussed further in Chapter 10.

or updated quickly, and hence at an early stage. A third strength is the way in which HIS can help to build public support for local policy, despite the fact that it has not yet been used in this way. However, it might add value since great importance is usually attached to information about health impact. Finally, the presentation of research results is also strong in HIS. HIS uses maps in which environmental nuisance and pollution are shown by means of contours or colour-shaded areas. Such maps are easy to understand and make it easy to visualize the situation, also for people without specialist knowledge.

However, HIS has a number of aspects that could be improved. The method is applied in full by the local municipal health service. When the screening is complete, the report is submitted to the commissioning party. There is virtually no interaction with stakeholders to qualify or interpret the outcomes. A second weakness relates to the comprehensiveness of HIS. Health impact is quantified, and this is only possible if measurement data are available. Where data are lacking, assumptions will be made, and this can distort the final picture. In the third place, the method is somewhat 'rough'. The various environmental aspects cannot be compared with each other in the HIS, apart from comparing HIS scores. It means that the scope for analysis is limited because it is only possible to compare scores. Finally, the method is optional not mandatory, which means that the commissioning party can decide not to use the screening results.

Despite its problems, HIS is an interesting method. After all, it makes it possible to obtain insight at an early stage into the health-related effect of plans. It is never too early to obtain such information. In fact, the later this information becomes available, the greater the time (and therefore money) required to amend a plan. HIS is one of the few methods that, in this way, explicitly introduces the theme of health into the spatial-planning process.

6.4 Finally: Checklist Methods in Perspective

This chapter shows that the checklist methods constitute a special category. They are not methods that work according to the input-output model: the information that is fed in does not produce a ready-made solution. Rather, they are methods that are used primarily to ensure that certain interests are not overlooked and are considered at the appropriate stage in the process. Indeed, they are designed to build-in a 'check' on whether environmental aspects are taken into account. This is most evident in Amsterdam's Environmental Performance System. The other two checklist methods also have an additional functionality. The Amsterdam *Milieuplaberum* is a process method for *structuring* the planning process,[27] and

27 Seen from this perspective, the *Milieuplaberum* shows similarities with the MIRUP approach (Section 7.3), which also offers an alternative planning process with a strong focus on environmental interests. However, MIRUP does not use a list of environmental measures.

Health Impact Screening by municipal health services results in clear maps and is thus strong in *presenting* the results for several purposes.

When the three approaches are compared with each other, additional differences come to light. The Environmental Performance System has an interesting history in terms of the attempts at anchoring in spatial-planning structures. This led to a confrontation with the Council of State. Hence, the system eventually helped to reveal how the Spatial Planning Act did not provide for the formal embedding of the method in spatial planning at the time. This was a disappointment because, when the method was used for the *IJ-oevers* zoning plan, it turned out to be an easy way to enhance a zoning plan with environmental aspects or measures. Moreover, the Amsterdam Environmental Performance System not only facilitates, but also creates standards. A negative score for a particular environmental measure requires extra effort in respect of other measures in order to achieve the minimum required level of environmental performance. This process is not encountered in the other methods.

The *Milieuplaberum*, a system of procedures and decision-making stages, does not express consequences. Instead, it guides the parties involved through the planning process and provides for the inclusion or linking-in of environmental measures at fixed stages. A similar argument applies to HIS: no consequences are linked to the results of the method. It is down to the commissioning local authority to decide whether and how the findings will be used. HIS is an excellent communication instrument for putting the environment on the political agenda and for underpinning environmental arguments.

All three checklist methods can be placed in the same quadrant of the framework (see Figures 6.1, 6.2 and 6.5). In Chapter 3, we related that position to the classic standards-oriented approach, whereby responsibility for directing and implementation rests mainly with the government/public sector. That is certainly the case for the checklist methods discussed in this chapter. Yet this need not be an obstacle to using such methods. On the contrary, due to their consistent focus on environmental interests, they can be excellent methods for managing the fragile relationship between spatial planning and the environment. It is likely that there will be an increasing need for this type of method in the environmental policy of the future. Spatial-planning issues are becoming increasingly complex and increasingly interwoven with other issues. In the hectic sphere of spatial planning, it is important to have methods available that can safeguard certain interests in a straightforward way.

The next chapter will focus on Quality Perspectives. These are methods that can be used to create solutions, taking account of the area-specific characteristics of the issue at hand. Here, too, checklists can be useful, for example by ensuring that certain aspects are not overlooked.[28]

---

28 The MILO project office of the VNG (Netherlands Association of Municipalities) advocates Health Impact Screening as a useful instrument for use in the MILO approach. Chapter 7 will focus on this method. Mention has already been made of the combination of HIS maps and Environmental Atlases (Chapter 8).

Chapter 7
Quality Perspectives: Towards Tailor-Made Area-Specific Solutions

In the Netherlands, area-specific environmental policy is not a recent phenomenon. Over the years, for large areas of the Netherlands, specific policies have been formulated with a view to realizing or maintaining a certain level of environmental quality. In the case of the Wadden Sea and the Veluwe national park, policy focuses on the 'green' environment, while the 'grey' environment is an important theme in the case of Schiphol. In many situations, spatial methods for the grey environment also make use of area-specific approaches.[1] The environmental zoning methods discussed in Chapter 5 can be regarded as area-specific policy instruments par excellence.

The mid-1990s saw the introduction of a new category of area-specific methods for the grey environment: Quality Perspectives. In this type of method, environmental quality targets are formulated as aims to be realized. The functions, possibilities and characteristics of the area in question play a crucial role in the formulation of these aims. This approach makes it possible to take account of the area-specific characteristics of the spatial project or issues in question.

The work method based on Quality Perspectives aligns well with decentralized environmental policy. Quality Perspectives pay greater attention to the potential of areas and the differences between areas. This means that there is room for specific policy *in addition* to generic policy. It is not only for environmental policy that these methods are a welcome addition. In recent years, they have also become increasingly relevant in spatial planning. With Quality Perspectives it is possible to coordinate environmental qualities and other relevant aspects of the physical environment (e.g. social and economic qualities) at area level. This enhances the coherence between the various policy sectors.

In this chapter we discuss three examples of Quality Perspectives. The examples selected are the 'Bandwidth Method' (Section 7.1), Environmental Quality Profiles (Section 7.2) and the Neighbourhood Target Scenarios (Section 7.3). The Bandwidth Method is the oldest of these methods and the Environmental Quality Profile is the most recently developed. The three methods will be compared in Section 7.4.

1 The study by Humblet and De Roo (1995), which has already been mentioned, is even presented as a study of area-specific environmental assessment methods.

7.1 'Bandwidth Method'

The Bandwidth Method was developed by a project group of the Regionaal Beraad Utrecht (RBU).[2] The aim of the project was to realize a methodology for the integrated assessment of the environmental input and performance of spatial plans. The activities began in 1994 with a grant from the Ministry of VROM, and the VINEX location Leidsche Rijn served as a test case for the first version of the methodology (BRU-projectgroep 1995). The first results were positively received. On that basis, the Bestuur Regio Utrecht (BRU) and the Ministry of VROM decided to make additional financial resources available for the further development of the method.

There are various reasons underlying the development of the Bandwidth Method. The first reason relates to the lack of instruments for assessing the contribution of the full range of environmental considerations to spatial plans perceived by the BRU (BRU-projectgroep 1996a, 1996b). The Bandwidth Method involves as many relevant aspects in the assessment as possible, thus covering a wider range of environmental considerations. A second reason relates to the availability of methods for assessing the role of the grey environment in spatial planning in the early 1990s. The methods available were mainly classic environmental-assessment methods (for a discussion, see Humblet & De Roo and Chapter 2). In planning the Leidsche Rijn urban expansion project, however, there was a need for 'an analysis methodology to facilitate the integrated consideration of aspects, and a presentation methodology to provide officials and citizens with clear information on the qualities of the plan' (BRU projectgroep 1996a; 2).

These reasons are reflected in the development of the methodology. Its application is thus intended to produce an integrated picture of the qualities of the plan in question (see Figure 7.1). In the quality picture, the scores for the various environmental aspects taken into account are visualized in the form of thermometers or histograms. The score for each aspect is presented on a scale that extends from a minimum value to a desired value. An interesting property of this instrument is that the effects are made visible for the supralocal environment, the local environment and in terms of the quality of the physical environment, instead of focussing only on the 'grey' environment (i.e. health and hygiene). According to the BRU project group, the methodology thus serves as a 'sustainability test for spatial plans' (1996a; 8)

Procedure

The concepts of 'sustainability' and 'sustainable spatial development' played an important role in the development of the methodology. Four principles have

2 The Regionaal Beraad Utrecht (RBU) is also known as the Bestuur Regio Utrecht (BRU; Utrecht Region Management). The BRU is an alliance of nine municipal authorities in the region of Utrecht.

been derived from these concepts. The development of the methodology involved translating these principles into themes: supralocal environmental quality, local environmental quality, quality of the physical environment and, finally, planning quality.

The methodology works as follows. Quality indicators are identified for each theme. These indicators must be relevant for the area in question, thereby facilitating an area-specific approach. In the case of the supralocal environment, for example, the indicators could be water use and energy use. The indicators for the theme 'local environment' relate primarily to environmental health and hygiene (noise, smell, air pollution, etc.). In the theme 'quality of the physical environment', the indicators could relate to the provision of amenities, the mix of functions, or public safety. Finally, the indicators for the theme 'planning quality' relate to the planning process (e.g. flexibility in phasing, or the involvement of residents and other stakeholders).

The indicators and related scores are shown in a graph or chart. A 'bandwidth' is set for each indicator (see Figure 7.1), which extends from the limit value (score 0) to the target value (score 100). The centre of the bandwidth (score 50) is regarded as an average level of ambition for the indicator in question. A score below 50 means that the quality of the indicator is 'reasonable'; a score above 50 is better. However, the plan is regarded as acceptable in terms of quality if the score is anywhere between 0 and 100.

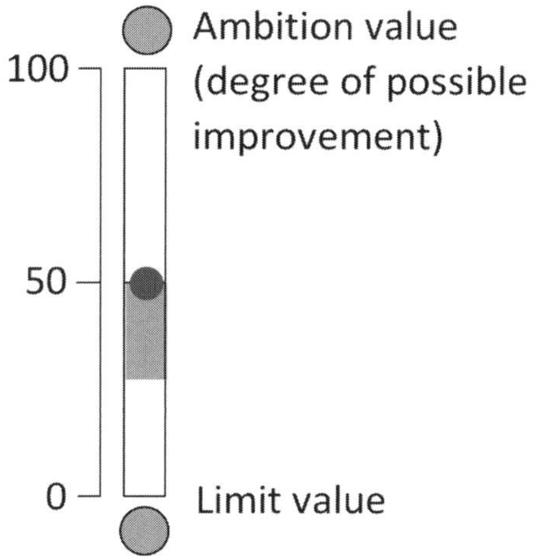

Figure 7.1 Indicator diagram (after BRU projectgroep 1996a)

An example shows how an indicator works. Suppose that, in the plan area, 8.4% of residents are subject to odour pollution. The maximum percentage of persons affected (the limit value) has been set at 12%. However, the aim is to reduce the nuisance to a level at which only 2% of residents are affected. The score for the indicator 'odour' can be determined from this information. The score is therefore: ((12 – 8.4) / (12 – 2) * 100 = 36. Here it must be noted that the percentage of persons experiencing nuisance is a question of definition, and the choice of limiting and target values is normative. This also means that the interpretation of quality must be carefully considered in advance. Without a satisfactory definition, the method will be little more than well-intended rhetoric.

This score in shown in the indicator diagram (Figure 7.1). In addition to the bandwidth, the diagram indicates the possibilities for improvement regarding the aspect in question, and indicates whether standards have been exceeded. This is done by means of a small circle at the top and bottom, respectively, of the indicator bar. An open circle means that there are hardly any options for improvement, or that no standards have been exceeded (i.e. there is little reason to act). A filled-in circle means that there are plenty of improvement options for the indicator in question, or that standards have been exceeded (i.e. there is much reason to act). Values between these two extremes are also possible.

This procedure is followed for every relevant indicator. Together, the indicators give an integrated picture of the quality of the plan. Because it focuses on quality, the method goes further than an examination of only functional allocation. This is very useful. The quality picture for the Leidsche Rijn is shown in Figure 7.2.

Practice

The formal development of the instrument began in 1994. Once the basic form of the instrument had been established, it was experimentally applied to the master plan for the Leidsche Rijn project. The report was published in 1995. The first application of the instrument was regarded as a success (BRU projectgroep 1996a), and its development continued.

The instrument was subsequently used by other municipal authorities. The Nieuwegein municipal authority has considerable experience with the instrument, which it used for the Blokhoeve and Galcopperzoom projects. These are both urban expansion projects based on sustainable spatial development. The project ambitions were calculated using the Bandwidth Method. In the case of the Blokhoeve project, the method was able to function in parallel with and supplement other methods – in this project, the DuBo-op-maat[3] and GreenCalc[4].

Discussion

The discussion of the Bandwidth Method has three parts. The first section assesses the position of the Bandwidth Method in Dutch environmental policy, based on the framework discussed in Chapter 3. The second part of the discussion involves

3 *DuBo-op-maat* is a performance-oriented decision-making model for weighing up the possibilities for sustainable building (SenterNovem 2007).

4 GreenCalc is a model for calculating and comparing the environmental impact of individual homes, buildings or neighbourhoods (SenterNovem 2007).

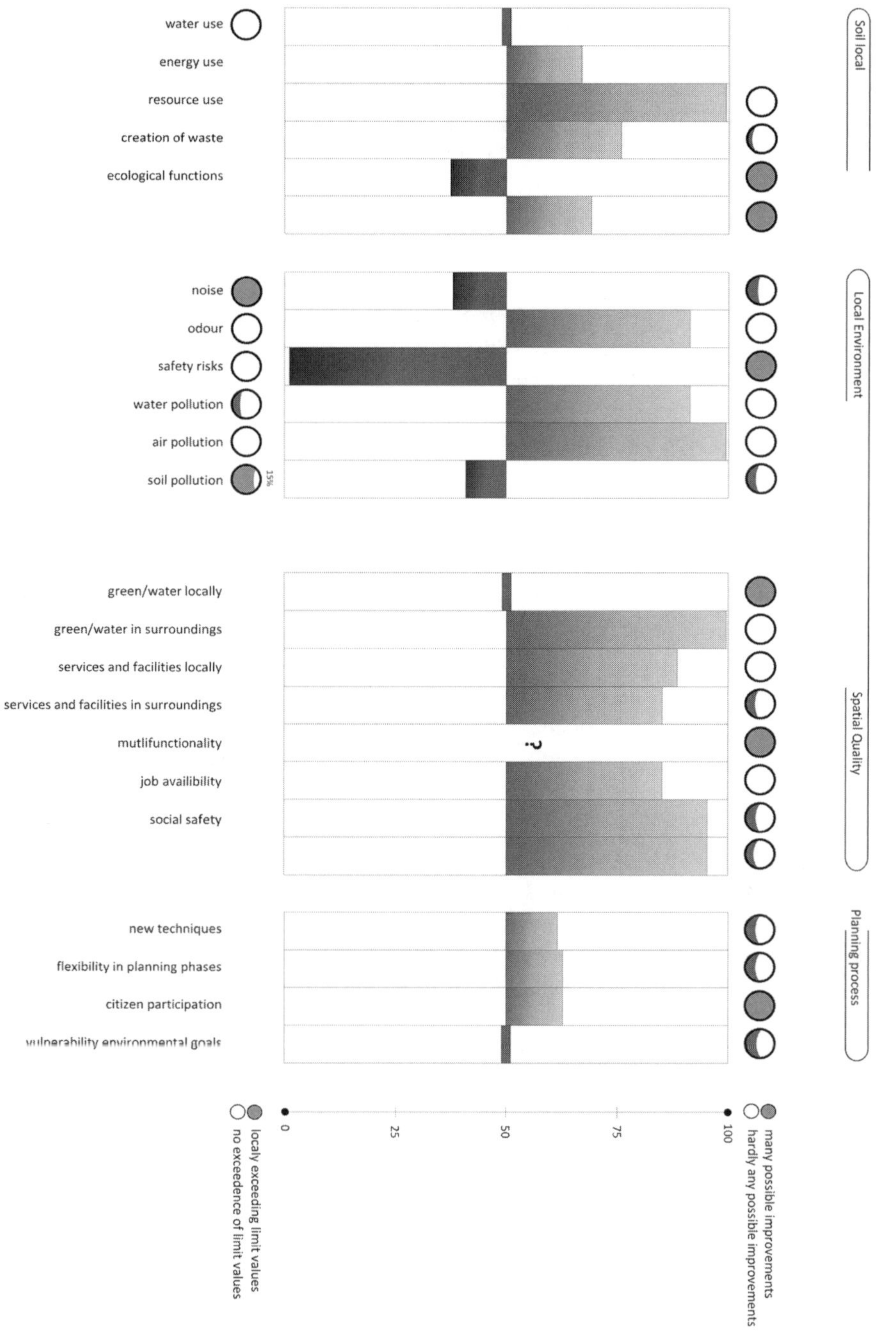

Figure 7.2 Integral quality picture for the Leidsche Rijn (BRU projectgroep 1996a)

a more in-depth analysis in which five characteristics of the Bandwidth Method will be examined. The third section focuses on the method's scope of application in planning processes.

One of the aims of the Bandwidth Method is to provide insight into the relationships between the various relevant environmental aspects in a spatial context. This results in an integrated quality picture. The relationship between the aspects, however, is assumed rather than demonstrated. This need not be a limitation, so long as it provides the user with clear information on the situation. This information is intended primarily for decision-makers and administrators. Therefore, there is little interaction with other actors. On the basis of these characteristics, it can be said that the method occupies a central position in the framework. The position is shown in Figure 7.3a.

The first characteristic to be considered is *intention*. This refers to the purpose of the methodology, as envisaged by its originators. The purpose is defined as providing insight into, among other things, the environmental qualities of spatial intentions. On the basis of this insight, administrators can weigh up the factors involved and make decisions. This means that, in terms of the characteristic *intention*, the method can be placed in the outer ring of the model.

The instrument considers not only local environmental quality (environmental health and hygiene aspects), but also the quality of the supralocal environment and the quality of the physical environment. In terms of the characteristic *integration*, the method can be positioned in the outer ring of the model.

We have already mentioned the fact that the methodology involves a limited amount of *interaction*. The main task of the initiator is to outline an integrated quality picture. The quality picture is presented in a strongly visual way, which means that the method is also suitable as a communication instrument. In practice, however, it is not always used as such. In terms of the characteristic *intention*, the method can be positioned in the middle ring of the model.

Area differentiation refers to the adaptability of the method in terms of area-specific characteristics. The Bandwidth Method provides for this by allowing the quality indicators to depend on the area in question. However, applying the method does not result in differentiation *within* the plan area, as is the case with zoning or other Quality Perspectives. In this respect, a combination of the Bandwidth Method and IMZ could prove very useful. In terms of the characteristic *differentiation*, the method can be positioned in the middle ring of the model.

The fifth characteristic is *conformance & performance*, which refers to the way in which the results of the method can be applied in the further planning process. The Bandwidth Method is primarily an analysis method. The result – an integrated quality picture – can provide grounds for reviewing certain aspects of a plan. The ultimate effect of the results, however, depends on how they are interpreted by decision-makers and administrators. This means that, in terms of the characteristic *conformance & performance*, the method can be placed in the middle ring of the model. Figure 7.3b depicts the relationships between the characteristics.

The third part of this discussion focuses on the applicability of the approach in the planning process. As stated above, the method is designed to 'work out' the quality of spatial plans. In order for the result to be utilizable in a plan, the method must be applied at an early stage. Only then is it possible to make adjustments. The way in which the quality picture is presented means that the method can also be used as a communication instrument. There are also possibilities for using it as a monitoring instrument. In 1998, an exploratory study was carried out for that purpose (Feenstra & De Wever 1998). Figure 7.3 gives a total picture of the aspects in this discussion.

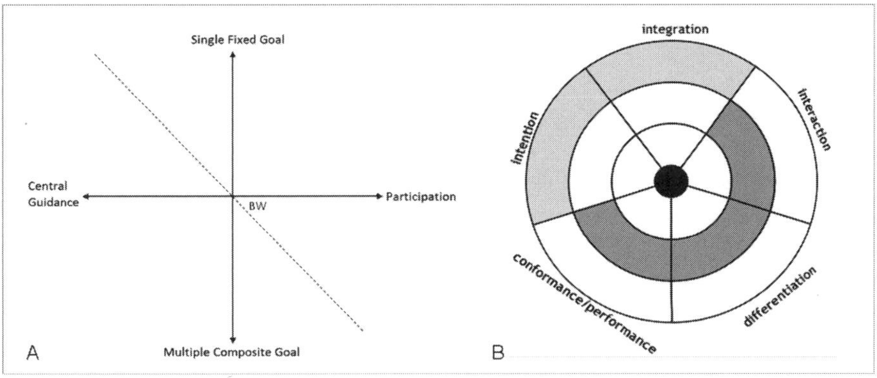

Figure 7.3 Characteristics of the Bandwidth Method

Conclusion

The survey reveals various strengths and weaknesses. The Bandwidth Method can be used in different phases of the planning process, and it is not necessary for previous phases to have been completed. It is designed for analyzing the relevant plan or planning process from an environmental perspective, and this can be done at any stage. Furthermore, the method focuses not only on the content-related aspects of the spatial plan. The quality of the planning process is also given attention and is an essential part of the analysis. In many other methods, this is not the case. The visualization of the integrated quality picture has been mentioned several times as one of the method's strong points. This means that it can also be used to inform stakeholders and the public about the quality of planned spatial interventions. Finally – and the Bandwidth Method is unique in this respect – it establishes a link between the area in question and the surrounding environment (the 'supralocal environmental qualities'). This provides information on the extent to which the plan lays claim to resources and ecological values outside the plan area.

A weakness of the method is the way in which the indicator score is calculated. This is sometimes done using very rough estimates and assumptions that are usually open to discussion. This means that the outcomes of the calculations – and hence the defined bandwidth – can be questioned. The application of the method also requires considerable input in terms of manpower and knowledge. If these resources are not available, there is a risk that the estimates for the indicators are little more than an arbitrary 'fudge'. Another weakness of the method is the way in which the indicator scores are graphically represented. At first glance, the presentation appears clear. However, doubts arise when the scores are examined more closely. The colours in the quality picture[5] give a misleading impression. They appear to suggest that a score below 50 (the average target score) is not a good thing. This is not the case, however.

Despite these reservations, the Bandwidth Method provides added value for large spatial projects. The applicability of the Bandwidth Method as a method for weighing up options and decision-making largely depends on the assumptions and preconditions (i.e. the 'bandwidth') that are defined in advance. It involves a relatively large number of assumptions, which compromise reliability and credibility.

7.2 Environmental Quality Profiles[6]

Most of the Methods for Environmental Externalities are 'stand-alone' methods that can be utilized in local environmental policy. However, there are exceptions. In Chapter 6, for example, we discussed the Environmental Performance System of the Amsterdam municipal authority. This methodology can be incorporated in the land-use plan. Methods from the 'Quality Perspectives' category can also be included in policy plans. A key example of this is the Environmental Quality Profile method of the Maastricht municipal authority.

In 2001, the authority presented its plan for nature and the environment (*Natuur- en milieuplan 2030;* NMP). The plan explicitly states that the environment is often taken into account too late in spatial planning processes. This can lead to problems at a later stage: decisions that have already been made may contravene environmental regulations. An important objective of the NMP was to firmly link abstract environmental objectives to concrete spatial plans (Gemeente Maastricht 2001). This was done using the Environmental Quality Profile instrument, which defines existing and desired environmental qualities for various area types and by specific environmental themes. The environmental qualities are then adapted to the spatial functions that occur in the area. This means that the quality levels aimed for will vary from area to area. For example, the permitted level of noise is higher on an industrial estate than in a quiet residential area. In the NMP this is described as aiming to realize 'various gradations of activity and quiet' (Gemeente

5 Green indicates that the aspect in question has a score above 50. Red indicates a score below 50. In general, the red is associated with a negative result.

6 The name Environmental Quality Profile sounds similar to the category 'Quality Perspectives', but it is a unique method.

Maastricht 2001; 17). In this way, suitable functions are allocated to areas, with a view to creating a high-quality physical environment (Ottens 2001).

Procedure

All methodologies that are regarded as Quality Perspectives have the same basic elements, i.e. the possibility for area differentiation, assessing the current qualities of an area, and defining the desired and required qualities. These elements are also found in the Environmental Quality Profile instrument of the Maastricht municipal authority. The procedure consists of three steps, each of which relates to one of the basic elements.

Area differentiation means that policy objectives vary according to the type of area in question. In a city centre, for example, the level of noise that is considered acceptable is higher than for a residential area. This procedure facilitates area-specific policy. In theory, the number of area types in this methodology is unlimited. However, it is advisable to keep to a manageable number of area types. The nature and environment plan (NMP, *Natuur- en milieuplan*) of the Maastricht municipal authority therefore uses eight types of area.[7] Criteria were formulated for each type of area.

It is possible that the plan area consists of only one type of area. This may be the case for a specific project, for example. However, several different types of area may be involved, as is the case in the nature and environment plan of the Maastricht local authority. The plan relates to the whole municipality. It is the initiator's task to formulate the criteria for the area types. In its NMP, the Maastricht local authority used four criteria to characterize the municipality. These relate to the main functions present, the mix of functions, intensity of use and the density of functions in the area (Gemeente Maastricht 2001).

Once the area types have been identified and allocated to the plan area or parts of it, the second step begins, i.e. formulating the desired qualities for the area. The Maastricht method first identifies quality levels[8] for each environmental aspect.[9] Similar qualitative distinctions are made for the various themes (Table 7.1). The quality levels are then allocated to the types of area. This can mean that the quality level 'moderate' is acceptable for the area type 'central environment', while only

7 The eight area types are: central environment, high-density living/working, low-density living, industry, low-density working, green connecting zones, outlying areas with landscape amenities and, finally, outlying areas (Gemeente Maastricht 2001). The last type of area (outlying area) has the character of a 'residual category' and is rarely used in practice.

8 With regard to setting the quality levels, the environmental plan of the Maastricht local authority stipulates that health-related limits must not be exceeded.

9 The initiator decides on the number of themes, as with the number of area types. The nature and environment plan of the Maastricht municipal authority uses seven types of theme: soil, water, disturbance, traffic, nature, waste and, finally, energy (Gemeente Maastricht 2001).

the quality level 'good' is acceptable for 'extensive living'. This distinction can give rise to unusual situations. It is possible, for example, that the quality level 'poor' is acceptable for an industrial park.

Table 7.1 Quality differentiation for the theme 'Noise' (Gemeente Maastricht 2001)

Quality	Description
Good	No persons experiencing noise nuisance
Satisfactory	Persons are affected by noise nuisance outdoors, not indoors
Moderate	Noise nuisance except at night, indoors
Poor	Noise nuisance indoors *and* outdoors

For each theme, the defined quality levels are linked to the previously defined area types. The resulting differentiated quality objectives are geared to the function (human use) of the area. This is known as the environmental quality profile. Table 7.2 shows the environmental quality profile for noise quality.

Table 7.2 Desired (noise) qualities for each area type (Gemeente Maastricht 2001)

Area type	Noise quality
Central environment	Satisfactory (general) Good (courtyards) Moderate (during events)
High-density living/working Low-density living	Satisfactory Good Satisfactory (access roads)
Industry	Poor
Low-density working	Moderate
Outlying areas of high landscape value	Good Satisfactory (access roads)
Green connecting zones	Satisfactory

The third step of the method involves determining the current quality levels in the various types of area. On this basis, it can be determined whether there is an 'ambition' (i.e. the current quality is below the desired quality) or whether the quality is satisfactory. This is also the stage at which measures can be suggested for achieving quality ambitions.

To summarize, the Environmental Quality Profile method consists of three steps. In the first step, area types are identified and allocated. In the second step, quality levels are determined for each environmental theme. The quality levels are then linked to the area types that have been identified. In the third and final step, the current quality levels in the area types are assessed. On this basis, environmental ambitions are defined.

Practice

In 2001, the Maastricht local authority linked this method to its nature and environment plan (NMP). Environmental Quality Profiles are also suitable as 'stand-alone' methods, however. Several other local authorities in the Netherlands also have experience with them, including the municipalities of Haarlemmermeer (2003), Hellevoetsluis (2002) and Zwijndrecht (2003).

Discussion

The discussion on Environmental Quality Profiles has three parts. The first section assesses the position of the approach in Dutch environmental policy, based on the framework discussed in Chapter 3. The second part of the discussion involves a more in-depth analysis in which five characteristics of Environmental Quality Profiles will be examined. The third section focuses on the method's scope of application in planning processes.

One of the most important characteristics of an Environmental Quality Profile is that it allows for area and quality differentiation. To a certain extent, this is a departure from the idea of uniform environmental standards for achieving a minimum basic level of quality, based on national standards. Instead, the profiles focus on identifying location-specific quality levels for the environment and liveability. Area-specific and realistic or feasible qualities are formulated for comparable and similar area types. The resulting Environmental Quality Profiles can be used in various ways. The NMP for the municipality of Maastricht, for example, is supplemented with these profiles in order to link, in a realistic area-specific way, environmental qualities and spatial circumstances. The method can also be used in discussions between officials, residents and developers on the desired environmental quality and the consequences for urban planning. On that basis, the method can be placed anywhere in the area from the centre to the lower right of the framework (Figure 7.1a).

This position is also evident when we consider the five characteristics. The first characteristic is *intention*, which refers to the purpose of the method as envisaged by its originators. The Environmental Quality Profiles are not a generic standard but, depending on the user, can serve as an area-specific and directional point of departure. In terms of the characteristic *intention*, the method can be positioned in the outer ring of the model.

The method, as applied by the Maastricht local authority, can also be used to identify various non-environmental qualities as relevant for an area or location. The Maastricht local authority works with seven themes, some of which are divided into sub-themes. The themes relate to environmental health and hygiene, as well as to aspects such as nature and public green spaces in cities. The quality profiles thus provide a far-reaching integrated perspective on an area or location. In terms of the characteristic *integration*, therefore, the method can be positioned in the outer ring of the model.

The first experiences with the environmental quality profile are based on official profiles. We have already mentioned the communicative uses of the instrument. An environmental quality profile can be used to inform residents and to discuss the desired level of quality with residents and other stakeholders. In terms of the characteristic *interaction*, the method can be positioned in the middle and outer rings of the model.

Area differentiation is identified as a core element of the instrument and requires no further explanation. In terms of the characteristic *differentiation*, the method can be positioned in the outer ring of the model.

The final characteristic is *conformance and performance*. In the introduction to this section, we mentioned the objective in the Maastricht NMP: to bridge the gap between spatial policy and environmental policy. Environmental quality profiles are incorporated in spatial planning policy at an early stage and therefore shape its development. In terms of the characteristic *conformance and performance*, the method can be positioned in the outer ring of the model. Figure 7.4b depicts the relationships between the characteristics.

The third part of this discussion focuses on the applicability of the approach. In order for Environmental Quality Profiles to form an effective bridge between spatial and environmental policy, they must be implemented at an early stage. Hence, they should be available at the beginning of the spatial planning process. In subsequent phases of the process they can be used for communication or evaluation purposes.

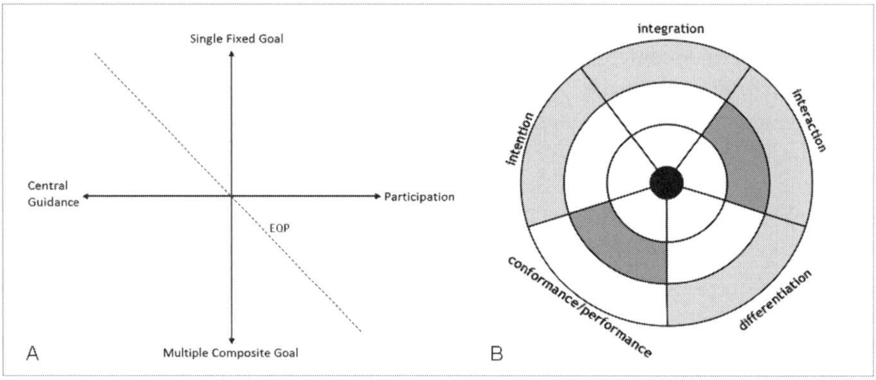

Figure 7.4 Characteristics of an Environmental Quality Profile

Conclusion

The survey reveals various strengths and weaknesses. It is interesting to observe that different types of area can be identified using more or less similar qualities. This process is geared towards a customized approach based on area differentiation. Refining this differentiation into area categories ensures that the diversity process remains manageable during the policy process. This customized, area-specific approach brings the policy fields of environment and spatial planning closer together. It also identifies the joint responsibilities (i.e. area qualities) of these policy fields. The area qualities serve as excellent input for communication and interaction with the residents involved. In fact, communication and interaction with stakeholders are likely to increase the support for policy geared towards the qualities. A final strength is the fact that the area qualities can also be visualized in the form of maps, which can reinforce the related discussion.

A weakness of the method is the extent to which the quality profiles are translated into concrete policy objectives. A profile comprises above all the policy ambitions of the environment sector, but does not specify how those ambitions will be realized. Another weakness of the method is the way in which the quality levels are substantiated. The description of the quality level 'moderate' (see Table 7.1) will not always correspond to what is generally accepted. A final weakness of the method relates to situations that are evaluated as 'poor quality' and then deemed acceptable. In an industrial park, the noise quality 'poor' is acceptable (Table 7.1). If this criterion is met in an industrial park, the quality is described as 'good' because it meets the level in the policy objective.[10] This approach can be confusing for outsiders.

Despite these reservations, the Environmental Quality Profile method is an interesting method for supporting local environmental policy. It allows environmental aspects to be introduced fairly easily into spatial planning processes. It is also a flexible method: area types and environmental themes can be added or omitted as required.

7.3 Neighbourhood Target Scenarios[11]

The municipality of Groningen is not unfamiliar with the phenomenon of methods for environmental externalities. In the early 1990s, the Environmental Service Department of Groningen developed the Groningen Environmental Assessment Method (Gemeente Groningen 1993). The aim of the method was to give

10 The example given here relates to the nature and environment plan of the Maastricht municipal authority. Obviously, the labels and descriptions of the various quality levels will vary according to the situation. This depends on the choices made by the initiator.

11 In Dutch, this method is called *Wijkstreefbeelden* or *Wijkgerichte milieustreefbeelden*.

environmental policy a proper place in local spatial planning. In order to realize this, an approach was chosen that would translate environmental quality in such a way that it aligned well with the typical jargon and culture of spatial planning. In the 1990s, the method proved its added value in Groningen's local environmental policy (for a discussion, see Humblet & De Roo 1995). In contrast to many other methods of the time, this method not only assessed the current situation, but also gave suggestions as to how the area in question could be structured.

However, as Dutch environmental policy continued to evolve (see Chapter 2), this assessment method became less usable. Decentralized environmental policy requires more communicative instruments that also allow for a wider range of interests and priorities to be considered. The environmental policy plan *Lokaal Gewogen 2001–2004* was adopted in 2001 (Gemeente Groningen 2001). It emphasized the importance of improving the quality of the physical environment and improving communication on the environment.[12] This plan laid the foundation for the later Neighbourhood Target Scenarios.

By way of experiment, in 2001 the Groningen municipal authority formulated target scenarios for two neighbourhoods: Oranjewijk and De Hunze. The scenarios describe the current and desired quality of the environment. They relate to noise, air, water and soil (which together form the 'grey' environment) as well as ecology, sustainability and a healthy indoor environment. The scenarios also contain an overview of possible areas for improvement. These points are identified on the basis of the current and desired qualities and in consultation with the public. The trial with the neighbourhood target scenarios was deemed successful (Gemeente Groningen 2001) and the municipal authority took the decision to develop them as a method for environmental externalities.

This was realized in 2002. The result is a method that the initiator can use to create a development framework (as an environmental obligation or environmental ambition) for a particular area. In addition, the method can be regarded as a test framework for proposed spatial developments, from an environmental perspective. The method is explicitly implemented in consultation with the local population. Residents are informed not only about the current condition of the physical environment, but also have a say in its desired quality. The instrument can therefore also be regarded as a communication instrument.

Procedure

When the method is applied, the outcome is a target scenario for the plan area. In the scenario, the environmental objectives and desired spatial developments are balanced against each other so as to find the most ideal or optimum mix of

---

12 In its next environmental policy plan, the Groningen municipal authority reported that the trial with Neighbourhood Target Scenarios had been successful (Gemeente Groningen 2005). The target scenarios were set out in 'neighbourhood environmental plans', and this approach was followed in the subsequent years.

qualities. The process for formulating a target scenario consists of three phases (Gemeente Groningen 2001).

The first phase is an integrated study of the basic qualities in the plan area. Statutory standards and existing municipal policy are the determining factors for the basic qualities (Gemeente Groningen 2003). In this phase, the municipal authority's ambitions for the area are also established. In relation to the basic qualities and ambitions, a distinction is made between environmental health and hygiene aspects (such as noise, waste, soil, air and water) and sustainability aspects. The latter relate to an aware and responsible use of space, resources and energy, and the preservation or reinforcement of ecological qualities. Close consultation between the services and departments involved in planning and managing the physical environment is crucial at this stage. Communication with the public does not yet take place in this phase.

The second phase of the process consists mainly of a dialogue with residents and stakeholders. This can be realized in different ways, for example by means of residents' evenings, complaint procedures or a liveability survey. Through these activities, the local authority informs residents and stakeholders about the basic qualities that have been determined and about its ambitions for the plan area. The parties involved should then indicate what their own ambitions are for the area. For example, they may want the plan to include more bus stops or cycle routes. These wishes can then be included in the target scenario. This is a process for finding out about the wishes of the public and other stakeholders for the plan area.

To summarize, the result of the first two phases consists of three elements. First, the basic qualities for the plan area are established on the basis of statutory standards and existing environmental policy. Second, the municipal authority sets out its ambitions. Third, residents and stakeholders make their ambitions known. In the third phase of the process, these three elements are combined into an integrated Neighbourhood Target Scenario (see Table 7.3).

A box shaded in black in Table 7.3 means that the indicator in question (e.g. road-traffic noise) does not comply with the basic quality. In the same way, a dark shaded area in one of the ambition columns means that there is a ambition relating to the indicator in question. A requirement exists for an indicator in the Neighbourhood Target Scenario if the box for the basic quality or one of the ambitions has dark shading. In Table 7.3, this is shown in black.

Neighbourhood Target Scenarios are formulated with reference to other municipal policy. In the case of Groningen, this means that they affect different visions for urban districts and the related implementation programmes (Gemeente Groningen 1993). There is a 'leapfrog' effect between neighbourhood scenarios and other municipal policy. This means that the most recent decision (the Neighbourhood Target Scenario) may contain amendments for earlier decisions.[13] However, since the neighbourhood scenarios have this status, they require the

13 Voogd (1996) questions the 'leapfrog' principle because of the absence of proper coordination between the different policy fields. It is actually more of a pragmatic solution.

Table 7.3 Neighbourhood Target Scenario for Hoogkerk (Gemeente Groningen 2001)

Hoogkerk		Basic quality	Ambition of local authority	Ambition of residents	Requirement in target scenario
Water	Quality of surface water				
	Quality of aquatic sediment				
Soil	Soil quality				
Air	Air quality				
Noise	Road-traffic noise				
	Railway noise				
	Industrial noise				
External safety	Businesses				
	Traffic routes				
Level of activity	Nuisance				
Public Space	BORG (Public-space management)				
Waste	Collection method				
	Underground containers				
Durability	Energy				
	Sustainable methods for building/ renovation				
Public transport	Distance between stops				
	Frequency				
Bicycle	Cycle routes				
Ecology	Ecological structure				
Quality of homes	Health				

	Satisfies basic quality threshold; no ambitions
	Does not satisfy basic quality threshold *or* ambition exists
	Requirement in Neighbourhood Target Scenario

formal approval of the municipal council. When the council has approved it, a neighbourhood target scenario becomes an official municipal policy document.

Practice

The Groningen neighbourhoods Oranjewijk and De Hunze were selected for pilot projects with neighbourhood scenarios. Experiences with the scenarios were positive (Gemeente Groningen 2001, 2003), and it was decided to draw up a scenario for every neighbourhood in the municipality. The information in Table 7.3 is from the target scenario for the Hoogkerk neighbourhood and is the basis for translation into the spatial-planning context (Figure 7.5).

The target-scenario method is now used in other parts of the Netherlands. For example, in the environmental policy plan for the municipality of Krimpen aan den IJssel it was presented as a framework for assessing spatial plans from an environmental perspective (DCMR 2003). The method was also used in Hellevoetsluis, where target scenarios were included in an environmental structure plan (Gemeente Hellevoetsluis 2002). The Apeldoorn municipal authority also did this (Arcadis 2001).

Discussion

The discussion of the target-scenario method has three parts. The first section assesses the position of the approach in Dutch environmental policy, based on the framework discussed in Chapter 3. The second part of the discussion involves a more in-depth analysis in which five characteristics of the target-scenario method will be examined. The third section focuses on the method's scope of application in planning processes.

One of the most notable aspects of the Neighbourhood Target Scenario instrument is its communication-oriented approach. In the first place, the instrument should facilitate internal communication within the local authority. In addition, there is extensive interaction with residents and other stakeholders. Another characteristic of quality profiles is that they work with basic quality thresholds and target qualities. On the basis of these characteristics, the method can be placed in the lower right section of the framework (Figure 7.6a).

This position also becomes evident when we consider the five characteristics. The first characteristic is *intention*, which refers to the purpose of the method as envisaged by its originators. In the case of Neighbourhood Target Scenarios, the originator's intention is to steer the spatial development of an area in such a way that environmental factors are considered. This means that, in terms of the characteristic *intention*, the method can be placed in the outer ring of the model.

One of the instrument's strengths is its breadth. It considers not only environmental health and hygiene aspects (e.g. noise nuisance or air pollution) but also establishes basic quality levels and ambitions for sustainability aspects.

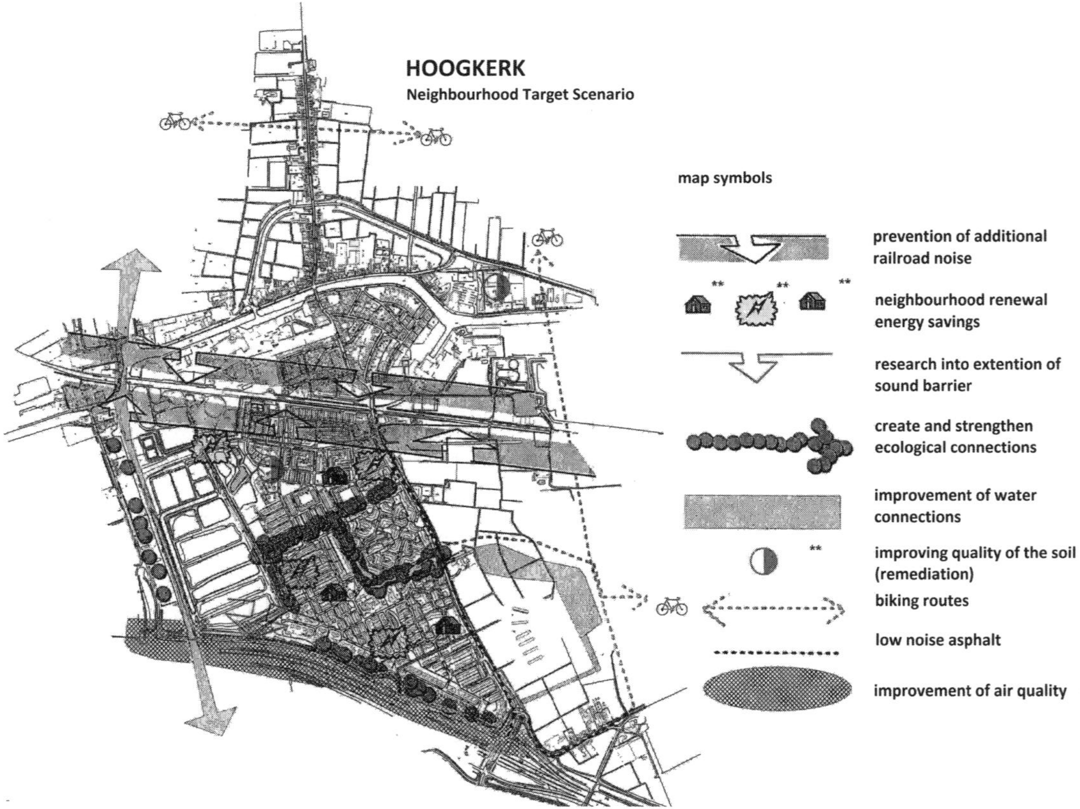

Figure 7.5 Spatial interpretation of the Neighbourhood Target Scenario for Hoogkerk

In terms of the characteristic *integration*, the method can be positioned in the outer ring of the model.

We have already mentioned the strong communicative nature of the instrument. In terms of the characteristic *interaction*, the method can be placed in the outer ring.

The possibility for area differentiation is a point of departure that can be found in all Quality Perspectives. Neighbourhood Target Scenarios are no exception. In terms of the characteristic *differentiation*, the method can be positioned in the outer ring of the model.

The final characteristic is *conformance and performance*. If an indicator does not meet the basic quality level, this results in an obligation in the target scenario. The same applies to the ambitions of the local authority and residents. However, no indication is given as to how or when this should be done. This means that, in terms of the characteristic *conformance and performance*, the method can be placed in the middle ring of the model. Figure 7.6b depicts the relationships between the characteristics.

A Neighbourhood Target Scenario expresses an ambition. Looked at closely, it involves points for improvements (the obligations in the target scenario) for the area in question. In order for these points to give direction to the process, the method must be used at an early stage in the planning process. In later phases of the process, the target scenario can be used as an evaluation framework. Figure 7.6 gives a total picture of the aspects in this discussion.

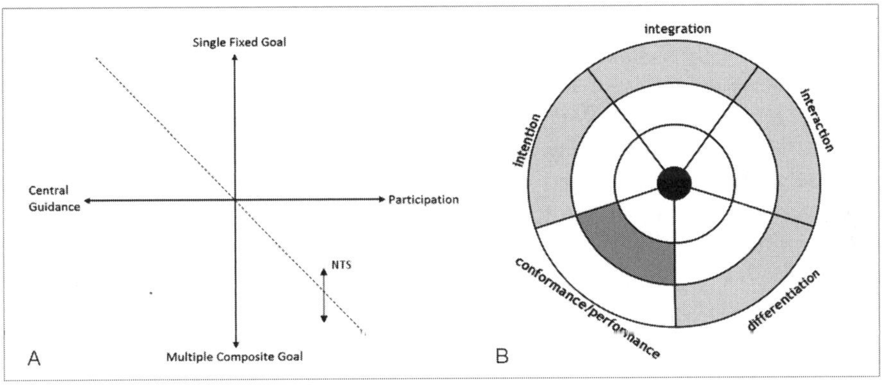

Figure 7.6 Characteristics of Neighbourhood Target Scenarios

Conclusion

The survey reveals various strengths and weaknesses. The communicative nature of the instrument has been mentioned as a strength. In the first phase, all relevant municipal departments are involved in the process. In the second phase, residents and other stakeholders are involved. Involving residents at an early stage provides a better picture of their wishes, and delays due to objections to policy proposals can be avoided. Another strength is the fact that the instrument can be used for newly created as well as existing situations. Finally, the accessibility of the instrument is also a strong point. The instrument uses clear overview tables and maps, which means that outsiders can also understand the situation.

A less strong aspect of the method is the fact that it does not encourage residents and other stakeholders to think ahead. The residents of an area are mainly occupied with the 'here and now'; the existing qualities of the environment. Moreover, residents usually think in terms of their own interests, rather than thinking of the collective good. Another weakness is the labour-intensive nature of the instrument. Organizing a communicative process between government and the public requires considerable time and effort.

Considered as a whole, the Neighbourhood Target Scenario is an instrument that aligns well with decentral environmental policy. Residents can see that their input helps to shape the target scenario. An area-specific approach in combination with basic and target quality levels (ambitions) appears to be a fruitful approach.

7.4 Quality Perspectives in Perspective

In recent years, policy practitioners have shown a great deal of interest in Quality Perspectives. The perspectives are methods that can be used to analyse existing and desired environmental qualities. Several different types of quality perspective are now used in the Netherlands. Although their approaches may differ, they are made up of the same elements: area types, quality indicators, references and ambitions. The Quality Perspectives discussed in this chapter can be used in various ways.

The Bandwidth Method of the BRU was developed as early as the beginning of the 1990s. It is therefore one of the oldest Quality Perspectives. The Bandwidth Method can be used to identify and discuss the effects of proposed changes. The effects for a number of indicators are given a score on a scale that extends from a minimum value to a target value. Options for improvements can also be indicated. One of the instrument's strong points, in which it also differs from the other two types of quality perspective, is that it identifies the effects on the supralocal environment as well as the quality of the local physical environment. The Bandwidth Method can therefore be regarded as an instrument for weighing up alternatives. This is also the case for the other two types of quality perspective. However, they differ in their approaches. In the case of Environmental Quality Profiles and Neighbourhood Target Scenarios, it is not a question of identifying the effects of intended interventions but of presenting the desired environmental qualities in an area profile, which then serves as the basis for the spatial-planning process.

The three instruments discussed each have great value in terms of communication. Quality Perspectives can be presented in a clear and straightforward way to the persons involved and to other stakeholders. Methodologies from the 'Quality Perspectives' category therefore align well with decentralized environmental policy. In such a decentralized policy setting, government authorities are more reliant on other actors, particularly with regard to policies for the physical environment. The scope for communication that the instruments offer – an aspect in which Groningen's Neighbourhood Target Scenarios excel – allows for joint discussions about minimum and desired quality ambitions. Despite this advantage, applying Quality Perspectives can be a time-consuming task. This is particularly the case with Neighbourhood Target Scenarios, in which residents and stakeholders are almost literally the 'designers'.

Finally, we can report that there is still a growing interest in Quality Perspectives. Practical experience with the perspectives is also encouraging other local authorities to consider the possibilities they offer in combination with local environmental policy. The Neighbourhood Target Scenarios developed in Groningen have now been used in several locations in the Netherlands. This is also the case for the Bandwidth Method and Environmental Quality Profiles.

The instruments discussed in this chapter are 'stand-alone' instruments. In the case of Environmental Quality Profiles, a link can be created with local environmental-policy plans. In recent years, new methodologies have been developed in which Quality Perspectives form part of a greater whole. The Environmental Agency for the Rijnmond region, for example, developed the LOGO instrument and the Stadsgewest Haaglanden (Hague Region) developed the MIRUP instrument. These are instruments that combine the possibilities of various methods for environmental externalities, including Quality Perspectives. Chapter 8 will focus on these instruments.

Chapter 8
'Zip' Methods: Integrating by Doing

The methods for environmental externalities discussed in the previous chapters were developed mostly to support a specific part of the spatial planning process or to resolve distinct sub-issues. To illustrate this, the purpose of Health Effect Screening is to identify the health effects of the intentions expressed in a plan at an early stage. The Environmental Performance System can be used to integrate environmental aspects in spatial plans, and the various Quality Perspectives can be used to formulate visions for development. These instruments can thus be a welcome *supplement* to policy for the physical residential environment. In recent years, however, methods have been developed that can be used to support the spatial-planning process from beginning to end. These are what we refer to here as 'Zip' methods.

Zip methods and Quality Perspectives have notable similarities. There is an important difference, however. The methods discussed in this chapter are largely based on an interaction-oriented approach, i.e. an approach whereby various departments and stakeholders are involved in the process. A total of three approaches will be discussed in this chapter. The first is the LOGO approach, which is discussed in Section 8.1. The tool was developed by the environment department for the Rijnmond region in order to improve the quality of the physical environment. Cooperation between the parties involved is an important feature of LOGO. Second, the MILO approach (Section 8.2.) was developed by the Ministry of Housing, Spatial Planning and the Environment (VROM). It provides a structure for defining the ambitions for the environmental quality of a particular area. The Hague Region (Stadsgewest Haaglanden) developed the third approach to be discussed: the MIRUP approach (Section 8.3). This approach can be regarded as a toolkit for the spatial planning process. The three methods will be compared in Section 8.4.

8.1 The LOGO Approach

LOGO (the Dutch acronym for Local Area Typology and Environmental Quality) originates in the Rijnmond region. In this highly urbanized area, in which there is also a high concentration of environmentally intrusive activities, environmental supervision is the responsibility of the DCMR Environmental Protection Agency. This involves activities such as issuing and checking environmental permits, registering and processing environment-related complaints made by residents, and providing information on current environmental quality in the region. Advising on policy is also a core activity of the DCMR.

Originally, the DCMR's policy advice was restricted to the 'grey' environment, with the key aspects being noise, soil, air and safety risks. The scope of this policy advice widened with the realization that integrated policies can in many cases be more effective than sectoral policies. In response, the agencies' focus shifted towards this more integrated type of approach to environmental aspects, spatial-planning aspects and other aspects that together influence the quality of the physical environment. The LOGO tool was developed within this wider framework.

LOGO provides a structure for defining area-specific ambitions for environmental quality. These are then linked to a spatial planning process. The purpose of LOGO is to align these environmental ambitions with the characteristics, functions and possibilities that exist in the area influenced by the spatial plan to be designed (DCMR 2004). In this respect, it is similar to the Quality Perspectives discussed in Chapter 7. The aim of Quality Perspectives is to draw up a profile of qualities. LOGO takes this a step further. It attempts actively to facilitate the introduction of environmental ambitions in an environmental programme or vision for environmental policy. It can also be used to formulate concrete implementation measures.

LOGO is intended for local authorities, the authorities that are usually responsible for the quality of the physical environment. The DCMR (2004), however, states that it can also be applied on levels other than the local level, albeit with modifications. For example, LOGO can be used to incorporate environmental ambitions in provincial environmental-policy plans. It can also be used when developing land-use plans, restructuring plans, district development plans and City & Environment projects.

Procedure

LOGO is built around seven distinct steps that are together intended to produce the area-specific environmental ambitions and facilitate their incorporation in spatial planning processes (Figure 8.1). Completing all the steps results in environmental ambitions for the areas involved, and the translation of these ambitions into other policies for the physical environment. LOGO can therefore be regarded as a process method: in addition to focusing on the substance of the issue, it structures the planning process (see for example DCMR 2004, Appendix 1). In doing so, the LOGO method strongly relies on cooperation: 'The key to the success of LOGO is optimum cooperation between the parties involved' (DCMR 2004; 7).

In the first two steps of LOGO, the emphasis is on analysing the plan area. The first step involves analysing the spatial functions and potential of the plan area. This can be done in several ways. The LOGO method uses the 'layer approach' (see VROM et al. 2004). The area is described in terms of three spatial layers: the subsurface, traffic networks and the current land use. In the second step, the area type is determined on the basis of the spatial-functional characteristics of the plan

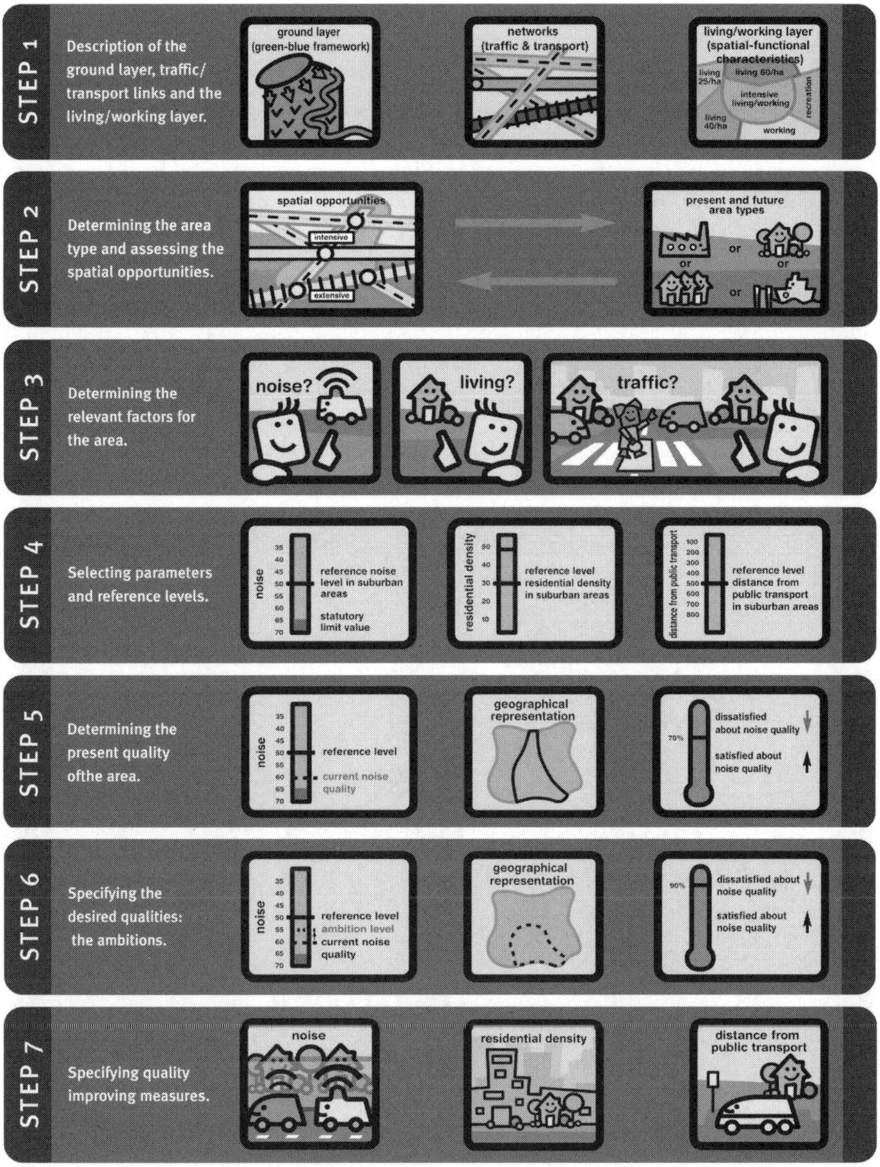

Figure 8.1 LOGO step-by-step plan (DCMR 2004)

area (i.e. the result of the first step). An area type is a label used for areas with similar general characteristics. These characteristics may relate to aspects such as the mix of land-use functions, the housing density and the intensity of use. Examples of area types include 'central environment', 'urban district' and 'leisure area'. It is possible to attribute more than one area type to the plan area. The user can choose from twenty-seven defined area types. If this number is not sufficient, the user can define and use a new area type. The area type is used in subsequent steps as a benchmark and to inspire the desired qualities.

Steps 3 to 6 involve determining the final environmental ambitions that are appropriate for the area type(s). In Step 3, relevant indicators are selected for expressing environmental quality in the area. The area type defined in the previous step is the basis for the indicators. Hence, safety risks will be a more important aspect in a city centre than in a nature-conservation area. Conversely, species diversity is more important in nature-conservation areas. In Step 4, quality parameters[1] and reference levels are then set for the selected quality indicators. The reference level for a quality indicator is shown in a bar chart (Figure 8.2). The reference level is the quality level deemed acceptable for the relevant indicator. Reference levels are intended only for measuring against current and desired quality. The statutory limit value can also be used for the indicator.

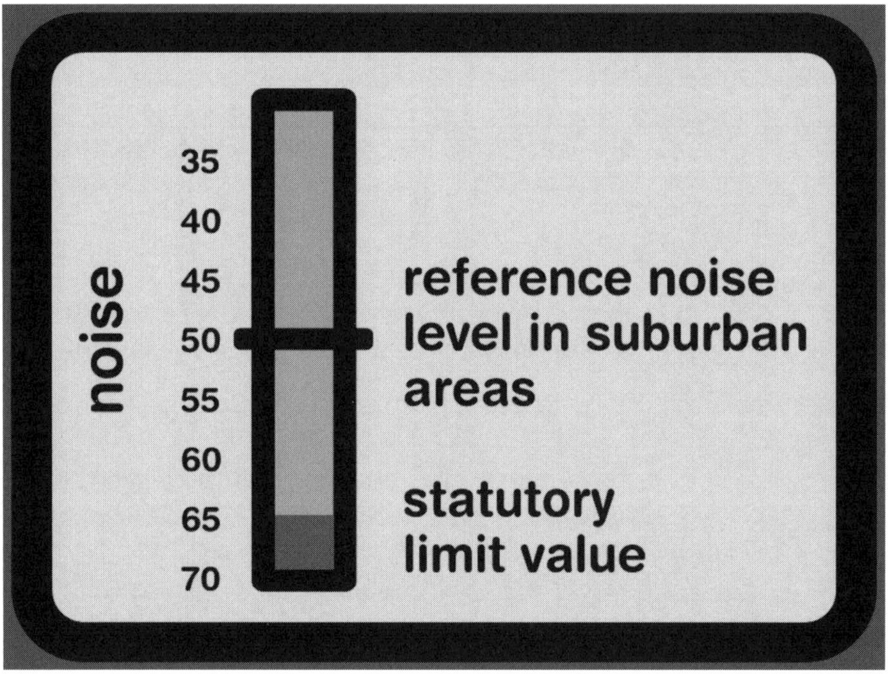

Figure 8.2 Example of reference levels in bar charts (DCMR 2004)

1 A quality parameter can be regarded as an entity. For example, if housing density is selected as a quality indicator for the area type 'suburban residential area', the quality parameter can be expressed as the number of homes per hectare. Air quality, for example, can be expressed as concentrations of particulate matter or nitric oxides.

In Step 5, the current qualities are determined for each indicator. These are then added to the bar chart (Figure 8.3). The reference level is always shown in the middle of the bar. If the quality allocated to an indicator is higher than the reference level, this is indicated with the colour green. Orange indicates a lower quality.[2] Red indicates that the quality exceeds the statutory limit value.

Step 6 involves determining the desired qualities for the selected indicators. Although the instrument does not specify whose responsibility this is, this step should obviously be carried out in consultation with all the parties involved. Together, the desired qualities constitute the environmental ambition for the area type. They are shown in bar charts, as are the reference levels and current qualities. The outcome of these six steps is a set of policy tasks: it becomes clear which types of improvement are necessary for each indicator (environmental aspect) (Figure 8.4).

Step 6 is not the final step in the LOGO procedure. The ambitions or policy tasks that have been formulated must also be put into practice. In the final step of the LOGO method, quality-improvement measures are formulated that should make it possible to realize the defined ambitions. The aim is to find creative solutions that can combine measures in such a way that different ambitions can be realized with a single solution; i.e. are thus integrated solutions. It gives the example of high-rise buildings that also serve as a sound bund for the buildings behind them. Another example is traffic measures that improve air quality as well as noise quality.

The seven steps of LOGO result in an integrated vision for environmental quality (Figure 8.5). This vision can be used in various ways in policy for the physical environment. It can be used to support a concrete spatial issue. An example is the restructuring of Stormeiland (DCMR 2006). The quality vision obtained with LOGO has been included in the master plan for this redevelopment project. It is also possible, as the method explains, to use the integrated vision to support the City & Environment approach. Another possible use is to identify tensions between the different functions in an area. The indicator bars facilitate this. Finally, the vision can be used to monitor the qualities in an area, without completing all steps of the plan. In such cases it is sufficient to measure current qualities against the desired qualities.

Practice

LOGO has been used in various projects, both within and outside the Rijnmond region. One example is the restructuring of the railway station area in Vlaardingen.

---

2 In this respect, LOGO is similar to the Bandwidth Method (Section 7.1). There is a difference, however. In the Bandwidth Method, the indicator bar is a *relative* indicator score. In LOGO, it expresses an absolute value. Furthermore, the 'zero point' in the Bandwidth Method is equal to the limit value. Consequently, if the limit value has been exceeded, it is not possible to measure by how much. This *is* possible with LOGO.

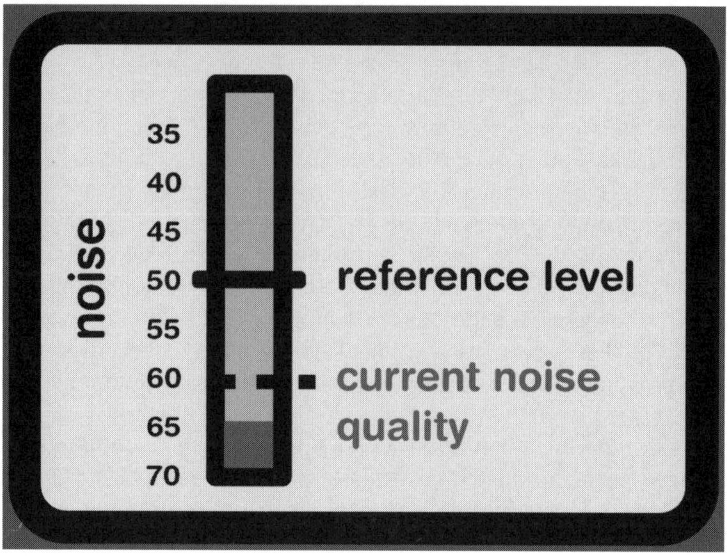

Figure 8.3 Example of current qualities as bar charts (DCMR 2004)

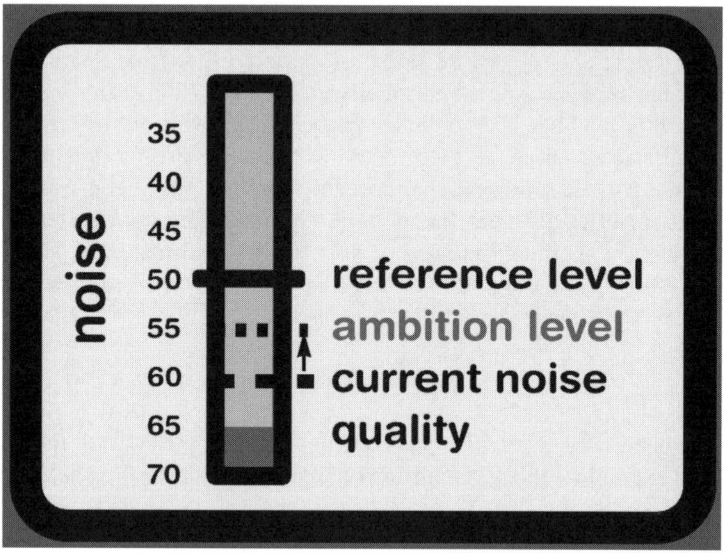

Figure 8.4 Example of a desired quality (ambition) in the bar chart (DCMR 2004)

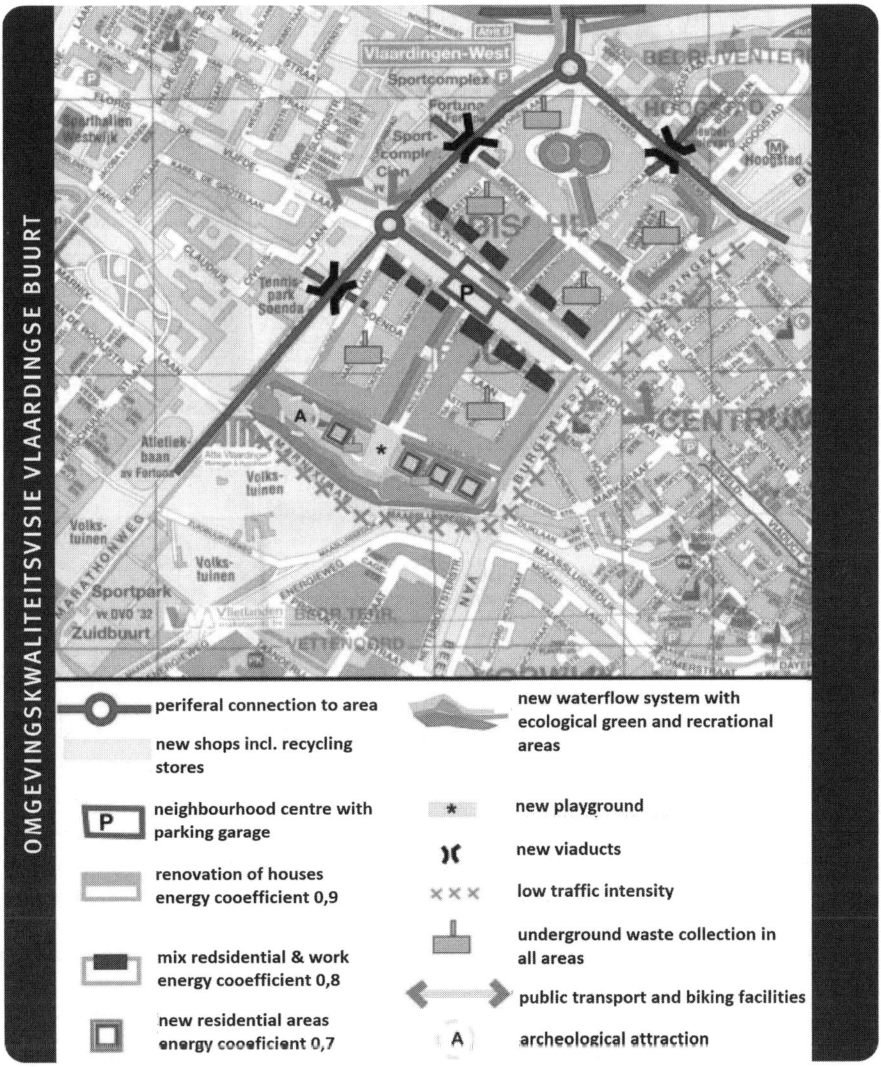

Figure 8.5 Integrated vision for environmental quality (DCMR 2004)

LOGO was used in this project because it allowed for the inclusion of environmental quality and sustainability aspects (in the form of ambitions) in the zoning plan. The method's added value was emphasized by those involved in the project. In January 2004 this process concluded with the adoption of the structure plan for the right bank of the river Maas (bron). This plan – including the area vision produced using LOGO – served as the planning and legal framework for further planning in the area.

In 2006, LOGO was used to support the restructuring project for an outdated industrial estate in the municipality of Krimpen aan den IJssel. The purpose of the restructuring project was to transform the old industrial estate into an attractive area for living and working. From the outset, the environmental department was involved in the development of a spatial structural concept for the area. LOGO played an important role in this. Parties involved in the project even regarded it as 'the ultimate example of early involvement of the sustainability sector in the development of a spatial structural concept' (interview with Huybregts 2006).

Discussion

The discussion of the LOGO approach has three parts. The first section assesses the position of the LOGO approach in Dutch environmental policy, based on the framework discussed in Chapter 3. The second part of the discussion involves a more in-depth analysis in which five characteristics of the LOGO approach will be examined. The third section focuses on the method's scope of application in planning processes.

There are several noteworthy aspects when assessing the LOGO approach. The method is oriented towards communication and participation: the involvement of stakeholders in the planning process is explicitly regarded as a success factor. Another notable aspect of LOGO is its flexibility. The user can choose from a wide range of area types that can help to identify the relevant indicators for the qualities in the area. The area types can therefore be seen as an a-priori set of indicators. However, users have the freedom to change the set; they make the choices. On the basis of these characteristics, the method can be placed in the lower right section of the framework (Figure 8.6a).

The method's position in the framework also becomes evident if we examine the five characteristics. The first characteristic is *intention*, which refers to the purpose of the method as envisaged by its originators. This can be defined as the systematic improvement of the quality of the physical environment, in cooperation with the relevant stakeholders. The method helps to structure the process, without imposing constraints. In terms of the characteristic *intention*, the method can thus be positioned in the outer ring of the model.

A second characteristic is *integration*, which refers to the extent to which the method considers environmental aspects in relation to each other. As this is one of the core ideas on which LOGO is based, it is no surprise that it can be positioned in the outer ring of the model.

This also applies to the characteristic *interaction*. LOGO refers frequently to the importance of cooperation between stakeholders. The procedure is geared towards this. In terms of the characteristic *interaction*, the method can be placed in the outer ring.

With twenty-seven different area types, the user has a wide choice. It is also possible to add to or change the set of area types. This means that, in terms of the characteristic *differentiation*, the method can be placed in the outer ring of the model.

The final characteristic is *conformance and performance*. The outcome of using LOGO is a set of environmental ambitions, which can be used to develop a vision for the area. The area vision can be used in the further structuring of the spatial-planning process. In terms of the characteristic *conformance and performance*, the method can be placed in the outer ring. Figure 8.6b depicts the relationships between the characteristics.

In order to ensure success, the LOGO tool must be applied at an early stage. It makes little sense to formulate environmental ambitions when certain decisions have already been taken. In the case of LOGO, this means that the method should be used in the initiative phase of a spatial-planning process. It can subsequently remain in place to play a role throughout the planning process. In later phases, for example, LOGO can be used to assess earlier decisions and monitor implementation.

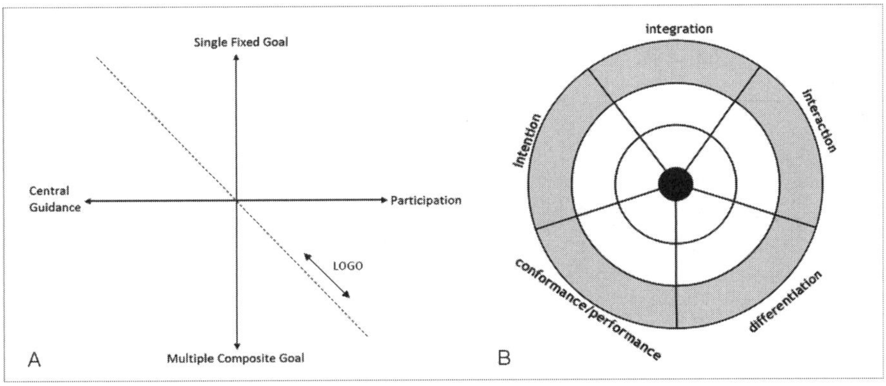

Figure 8.6 Characteristics of LOGO

Conclusion

The survey reveals various strengths and weaknesses. A strong point is the imagery that is used. This common and accessible language means that actors from diverse backgrounds can understand each other more easily. The area vision links the qualities and functions of the area. This aligns with the 'language' of urban planners and spatial planners, thereby facilitating cooperation between the disciplines. The images used in LOGO also make it easier for residents to understand the quality of the physical environment. Another strength of LOGO is the fact that environmental aspects are considered at an early stage when spatial issues are being resolved. This proved particularly valuable during the Stormeiland restructuring project.

LOGO focuses on the step-by-step procedure and the way in which an integrated area vision can be realized. The way in which the area vision is anchored in other local policy fields is at least as important. However, LOGO appears to underemphasize this. Process recommendations and success factors are not mentioned until the appendices, giving the impression that they are of secondary importance. The way in which the 'traffic-light' columns are used can also be questioned. In these columns, green indicates the amount above the reference value, and orange indicates the amount below it (see also Figure 8.3). Everything below the statutory limit value is coloured red. This does not allow for the fact that, in certain circumstances, there may be absolute limits. This may be the case, for example, when public health is at stake, and the current standards take the form of circulars or guidelines (i.e. extralegal). Such a situation may be in the orange area rather then the red area. This can give a distorted picture for decision-making.

Nevertheless, all things considered, the LOGO method fits very well in today's environmental policy. Practical experience with the method shows that it creates the link between the environment and spatial planning that is so necessary.

8.2 MILO Procedure

At a symposium in May 2004, MILO (the Dutch acronym for Environmental Quality in the Physical Environment) was presented to environmental officials and spatial planners from municipal authorities, environmental health departments and water boards. MILO was the result of a joint project involving the Ministry of Housing, Spatial Planning and the Environment (VROM), the Interprovincial Council (IPO) and the Association of Water Boards (UVW). MILO provides a concrete basis for an area-specific approach to the quality of the physical environment. This is also confirmed in the general guidebook for MILO, which states that the formulation of specific environmental ambitions for areas is central (VNG et al. 2004; 2). MILO thus has striking similarities with the LOGO instrument developed by the DCMR [3]

The main purpose of MILO is to formulate environmental ambitions for areas. These environmental ambitions should be aligned with the area's spatial functions, characteristics and possibilities. The underlying idea is that this approach provides optimum protection for the area's qualities. Originally, MILO was designed for local and regional authorities to use when formulating policies for the physical environment. In addition, the MILO project bureau was set up in 2004 under the Netherlands Association of Municipalities (VNG), to support these authorities in the implementation of the MILO instrument.

Soon after the presentation of MILO, it became apparent that the method was too inflexible. Municipal authorities that used the method felt that it offered too few possibilities (interview with Van Helden 2006). It was also found to be too complex to use. Gradually, on the basis of these experiences, the method has been modified. The MILO project bureau has taken the lead in this. Today, MILO is no longer a 'method' as such; rather, it is now seen as the 'MILO procedure'.

---

3 These similarities are not unexpected. LOGO (DCMR 2004) was the model on which MILO was based.

The MILO procedure no longer involves the original 'seven steps' (which were very similar to the LOGO steps). Instead, it combines various approaches for integrating environmental considerations in spatial plans, without the need to follow a distinct step-by-step plan. These approaches are partly inspired by the various steps, but are now considered as stand-alone approaches. The MILO procedure is therefore mainly a 'toolkit' from which the appropriate tools can be selected in each case. The following description of the method relates to this modified MILO approach.

Procedure

In the new-style MILO procedure there is more than one way to create integrated area-specific solutions. Selecting a procedure depends, on the one hand, on the type of issue faced (realizing a new residential area requires a different approach than a restructuring project). On the other hand, the stakeholders' wishes with regard to the planning process are a second dependent factor. The latter aspect is particularly important: the active involvement of all stakeholders in an open planning process is seen as a crucial success factor.

The new MILO procedure is built around three main components: (1) the analysis of the plan area, (2) the formulating of quality perspectives and (3) the provision of 'handles' for the spatial planning process. It is not necessary to use all three components. If a local authority only requires methods and techniques for drawing up quality profiles, that component of MILO will suffice. MILO thus provides instruments for various phases in the spatial planning process.

With regard to the first component, *analysing the plan area*, MILO provides three concrete instruments. The first of these is the 'layer approach' (see VROM et al. 20004) for assessing the spatial structures of the plan area. The second instrument is the environmental study. In contrast to the layer approach, a study is specifically geared to assessing the current environmental circumstances in the area. This may involve circumstance-related aspects of the 'grey' environment, but also the use of sustainable energy, forms of water storage or public green spaces. Much like the LOGO method described above, the current quality for each relevant aspect of the area or project is described. In addition, relating to this, the relevant legislation and standards are identified for the environmental aspect(s) as a benchmark. A third instrument that can be used to analyse the plan area is Health Effect Screening (see also Chapter 6). While HES can be used as a separate analysis instrument for the plan area, it is also possible to use it as an environmental survey in MILO. Similar to its description in Chapter 6, it gives an indication of both pollution and the presence of environmentally sensitive functions. Often, as MILO suggests, in cases where both exist it is advisable to use HES in order to translate the environmental situation into consequences for public health.

With regard to the second component, formulating environmental ambitions (Quality Perspectives), the possibilities for improving environmental quality in the plan area are identified. The underlying principle is that not all areas are the same. Consequently, the quality-related aims will vary from area to area, while the possibilities for improving the area will also differ. The process for formulating quality ambitions is explained in detail in Chapter 7 and is similar to the one that MILO proposes. For one thing, there must be sufficient information on the plan area (obtained in the first step of MILO). This information is then used to define the area type, in a similar way as described earlier regarding the LOGO approach. Thirdly, then, the area type is used to determine the desired qualities, i.e. the 'environmental ambitions'. In doing so, the MILO procedure aims to stay close to the procedure as discussed for the Environmental Quality Profile method (see Section 7.2).

Finally, MILO is used to structure the *spatial planning process*. In MILO, the planning process is regarded as comprising five main phases. These include the two components of the MILO procedure just described, i.e. making a study of the plan area and defining environmental ambitions. Each of the five phases begins with perspective-widening (gathering information), then focuses on the relevant aspects, makes choices and sets priorities (Figure 8.7). Each phase can be concluded with a product (Initial Policy Document, Schedule of Requirements, zoning plan, etc.). At various stages in the planning process, alternative methods for environmental externalities are indicated that can be used to support the relevant phase of the process. Consequently, MILO also provides a structure that links existing methods for environmental externalities. Some of these methods are discussed in this book.

The updated MILO approach described above has few similarities to the original MILO. The latter was developed primarily to help local authorities to formulate environmental ambitions (quality perspectives). In addition, it gives indications and tips for the spatial planning process. In the new MILO approach,

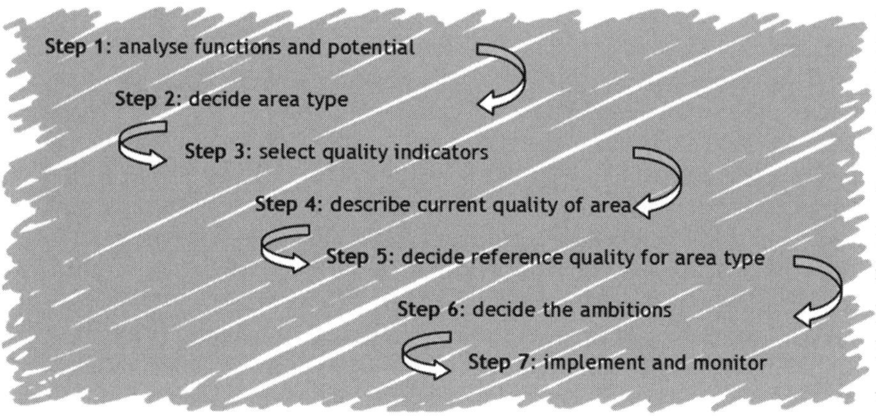

Figure 8.7 Diagram of a planning process according to the MILO approach (VNG et al. 2004)

this approach is in fact reversed. The spatial planning process is taken as the starting point. The focus is continually on integrating environmental aspects in the process and safeguarding environmental interests.

Practice

The MILO Project Bureau was set up to assist local authorities when the MILO method was introduced. The method has been widely used, albeit mostly in its updated form. A project overview drawn up by the MILO Project Bureau can be found on the website of the VNG (Netherlands Association of Municipalities). In the vast majority of cases, one or more components of the method were used. In only a few cases were all components used.

Discussion

The discussion has three parts. The first section assesses the position of the MILO approach in Dutch environmental policy, based on the framework discussed in Chapter 3. The second part of the discussion involves a more in-depth analysis in which five characteristics of the MILO approach will be examined. The final section focuses on the method's scope of application in planning processes.

Although MILO is presented as a procedure to ensure the optimum integration of environmental considerations in spatial planning processes, it mainly offers a structure for linking various methods for environmental externalities to spatial policy-making and implementation. Since MILO brings together various methods, it is difficult to position it clearly in the framework. Nevertheless, considering its points of departure makes this possible to a certain extent. One main point of departure is that MILO invites all policy sectors to take part in the process. Close cooperation is seen as a success factor. Another important point is that the environment is seen as an entity that can be weighed against other factors. This is reflected in the use and area types and quality perspectives. On the basis of these characteristics, the method can be placed in the lower right section of the framework (Figure 8.8a).

The first characteristic to be considered is *intention*. This refers to the purpose of the methodology, as envisaged by its originators. MILO is designed to integrate environmental requirements in spatial plans. To this end it provides a structure and instruments for the spatial planning process. The users of MILO decide which instruments to use and how to apply them. Because of this flexibility, the method can be positioned in the outer ring of the model.

MILO offers sufficient scope to include all relevant environmental aspects in the process. This is actually a core element of the method. In terms of the characteristic *integration*, the method can thus be positioned in the outer ring of the model.

The MILO guidebook is explicit in its reference to cooperation between the different policy sectors. Communication and cooperation are regarded as essential to the success of the process. In terms of the characteristic *interaction*, the method can be positioned in the outer ring of the model.

One of the MILO instruments is designed for formulating area-specific environmental ambitions, thus providing for area differentiation. In terms of the characteristic *differentiation*, the method can be positioned in the outer ring of the model.

The final characteristic is *conformance and performance*. The ultimate aim of the MILO approach is a spatial plan that gives ample consideration to environmental aspects. In part, the formulated environmental quality ambitions are even informing spatial planning choices. However, they do so in a inspirational and informative way. Hence, in terms of the characteristic *conformance and performance*, the method can be positioned in the outer ring of the model. Figure 8.8b depicts the relationships between the characteristics.

The third part of this discussion focuses on the applicability of the MILO approach in the spatial planning process. MILO is built around various components, the use of which is not obligatory or always necessary. Therefore, users of the methods are free to select the elements from the MILO approach that best suit their purposes. In the meantime, this flexibility is supported by the structure for the spatial planning process that MILO offers. 'Handles' are then provided for all phases of the spatial planning process, and these can be used to monitor environmental considerations. This means that MILO is suitable for all phases. This requires further explanation. Because the method builds on the spatial planning process, the effects will depend on the phase in which the method is used. The components of MILO (i.e. analysis, quality ambitions and structuring of the spatial planning process) are all designed to support decision-making. The strength of the method therefore lies in its input up to the decision-making phase. This input can be secured in the final phase.

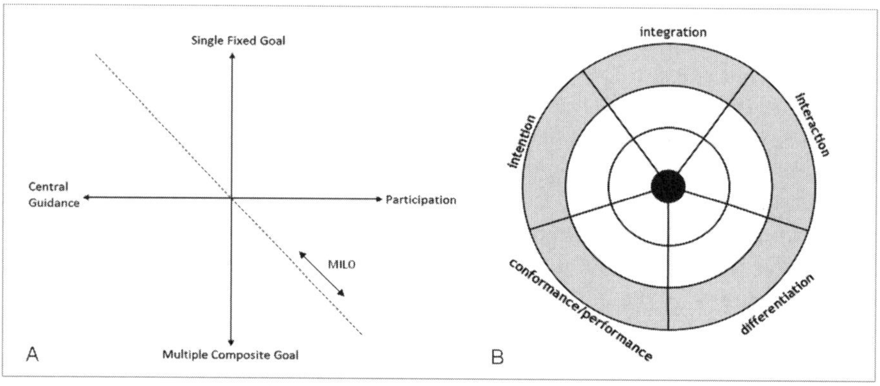

Figure 8.8 Characteristics of the MILO method

Conclusion

The survey reveals various strengths and weaknesses. One of the MILO method's strong points is that it brings together existing methods for environmental externalities. For the study of the plan area, for example, three instruments are taken as an example. Depending on the issue, the user can select one or more of these. Another strength is the structure that the method provides for the spatial planning process. In doing so, the appropriate support instruments are also indicated for each phase. Central to the method is the idea that environmental quality ambitions are always intended as input and criteria for spatial decision making. MILO thus aligns with a trend in decentralized environmental policy whereby the environment is seen as an entity that can be weighed against other aspects but must not be ignored.

Local authorities using the MILO method were supported by the MILO Project Bureau, which has extensive practical experience. The resulting information is now managed by the VNG. Initially, the project bureau was set up for a period of two years. The period was extended but has in the meantime elapsed. Now that the bureau no longer exists, local authorities will have to take the initiative to anchor or utilize existing experiences. This can lead to problematic situations: MILO's many possibilities and instruments mean that it is not a straightforward method. Another of the method's weaknesses is the fact that it is time-consuming. It takes time to study the plan area and the spatial issue, and to formulate environmental ambitions. And these are only two phases of the planning process.

8.3 The MIRUP Method

MIRUP was developed in 2004 on the initiative of the Hague Region (Stadsgewest Haaglanden). MIRUP is the Dutch acronym for *Milieu in ruimtelijke plannen* ('Environment in Spatial Plans'). The name reflects the basic goal of the method: the instrument 'was developed to accord the environment a better place in spatial plans (Stadsgewest Haaglanden 2004; 5).

Cooperation between stakeholders is an important point of departure in MIRUP. It builds on the practical experience of the fact that the lack of cooperation/ coordination between local authority departments can cause problems. In some cases, the material contribution of environmental departments does not go beyond setting conditions or retrospective testing of spatial interventions against environmental health and hygiene standards. This is why spatial planning departments often perceive environmental policies as restrictive and problematic. One of the objectives of MIRUP is to have departments work together from the outset, so as to ensure that environmental issues and ambitions remain on the agenda throughout the planning process.

A further objective of MIRUP is to place the local issue and its possible solutions in a broader perspective. For this purpose, a distinction is made between vertical and horizontal perspective-widening. In the case of perspective-widening in the horizontal sense, the local issue is placed in a regional perspective: 'the [plan] area must fit within the municipal territory as a whole' (Stadsgewest Haaglanden 2004; 5). As the vertical horizon is widened, the relationship between the spatial task and the relevant environmental aspects is addressed. Either way, the widening of perspectives is intended to be a point for attention throughout the planning process. Only then, as the argument goes, can environmental aspects really contribute proactively to solving local planning issues.

Certain components of LOGO and MILO can be found in MIRUP. They include various area-analysis techniques and the use of area types and quality perspectives. There is a difference in emphasis, however. The LOGO and MILO approaches focus mainly on content, while MIRUP focuses mainly on the process of planning. MIRUP comprises different instruments for this purpose. One of the instruments is designed to provide 'handles' for the spatial planning process itself. The other instruments are designed to incorporate environmental considerations at various stages in the planning process.

Procedure

The procedure is explained in terms of the spatial planning process described in the method. This process is referred to further as the MIRUP planning process. The method was developed to allow proactive environmental input when realizing spatial tasks. For this reason, the MIRUP planning process is presented as facilitating and not mandatory (Stadsgewest Haaglanden 2004). It offers points of departure for both the material (objectives) and process-related aspects of planning. Ultimately it is the stakeholders who decide, in mutual consultation, how the process will be implemented. This means that there is scope for ideas and solutions tailored to local circumstances.

The MIRUP planning process can be divided into four phases. Each phase consists of a number of steps. The MIRUP planning process is illustrated in Figure 8.9. The consecutive phases are Study, Programming, Design and Development. MIRUP pays little attention to the final phase (Development). The individual phases need not follow on from each other. Phases can overlap, and certain phases may be completed more than once before a satisfactory result is obtained. One or more of the MIRUP instruments can then be used, depending on the nature of the issue in the relevant phase. Other methods and techniques can also be used in addition to these instruments.

The purpose of the Study phase (Figure 8.10) is to identify the spatial issue, the structures of/in the plan area and the relationship between the plan area and surrounding areas, and to identify the stakeholders. This phase therefore involves collecting and interpreting factual information. As far as possible, solutions for the spatial issue should not yet be considered. The Study phase is instead intended only to provide insight into the opportunities and preconditions for sustainable spatial development. The end product of this phase is Terms of Reference that

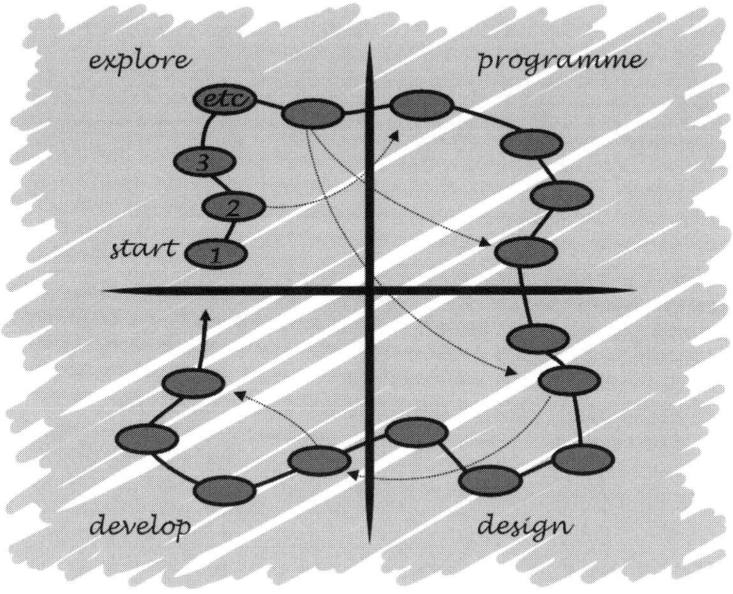

Figure 8.9 The four phases of the MIRUP planning process (Stadsgewest Haaglanden 2003)

set out the opportunities and starting points. MIRUP proposes the layer approach (see VROM 2004c)[4] as an instrument for structuring the Study phase. Other methods and techniques can also provide added value. For example, Health Effect Screening (Chapter 6) can be used to identify environmental aspects that constitute a potential threat to public health.

The second phase involves *programming.* The purpose of this phase is to establish a spatial structure for the plan area that incorporates the opportunities and preconditions identified in the Study phase that contribute to a sustainable spatial development. The spatial structure specifies, among other things, which environmental themes will be addressed. Target scenarios are also drawn up. These are based on the area type identified for the planning area. Two instruments are suggested for use during this programming phase. They are very similar to the Quality Perspectives discussed in Chapter 7. The first instrument can be used to define the area type. The second instrument is designed for drawing up target scenarios. The result of this phase is a set of environmental ambitions that is tailored to the spatial issue. These are set down in what is called a 'Schedule of Requirements'.

4 The 'layer approach' is a model for describing, analysing and explaining the spatial area. The approach is explained in detail on the website Ruimte x Milieu (http://www.ruimtexmilieu.nl).

The third phase (*Design phase*) involves the translation of the Schedule of Requirements into the final spatial plan. The environmental ambitions formulated in the second phase are used as conditional guidelines. The purpose of the Design phase is to produce a commonly accepted spatial plan that fully respects and incorporates the environmental ambitions defined earlier. During the Design process, as in the preceding phases, there should be close cooperation between the stakeholders. A catalogue of measures[5] for realizing various environmental ambitions is provided as a support instrument for this phase. Stakeholders can use the catalogue for ideas or select measures directly from it. The MIRUP process ends when the spatial plan has been produced. The method is designed to incorporate environmental aspects in the spatial planning process and, once the spatial plan is complete, this goal has been achieved. But this is not the end of the planning process.

The final phase is the *Development phase*. During this phase, the spatial plan – which has been officially approved – is implemented. In order for implementation to succeed, implications for other policy sectors and public bodies must be considered in advance. Similarly, the designed plan has to be translated into concrete projects and measures. MIRUP does not provide a structure for this phase; the MIRUP process ends when the spatial plan has been drawn up. MIRUP only provides some of points for attention in anticipation of the Development phase.

The four-step MIRUP planning process just explained is the core of the method. The process provides handles for introducing environmental aspects into spatial planning processes. Furthermore, this input is monitored throughout the process. At the appropriate stages, indications are given as to which steps can be taken and which instruments can be used. Notably, introducing environmental aspects into the spatial planning process is regarded as including sustainability ambitions in the spatial plan. The terms 'environment' and 'sustainability' are often used interchangeably. This is not only strange but can also be confusing.

It seems that the method is designed primarily to inspire its users. The various phases and steps in the planning process are explained using images, so that actors from diverse disciplines can communicate with each other more easily. In addition, the steps are illustrated with many examples.

Practice

MIRUP was developed for municipalities in the Hague region (Stadsgewest Haaglanden). The area types are therefore defined on the basis of the region's characteristics. But this does not mean that MIRUP cannot be used in other regions. On the contrary; the area types are presented as a guideline and not as blueprints. Area types can always be adapted.

---

5 The measures are organized by theme and linked to the area types. Examples of measures for the theme of 'Noise' in the area type 'Urban area' include fountains as a sound bund, office buildings to shield homes, and restriction of traffic.

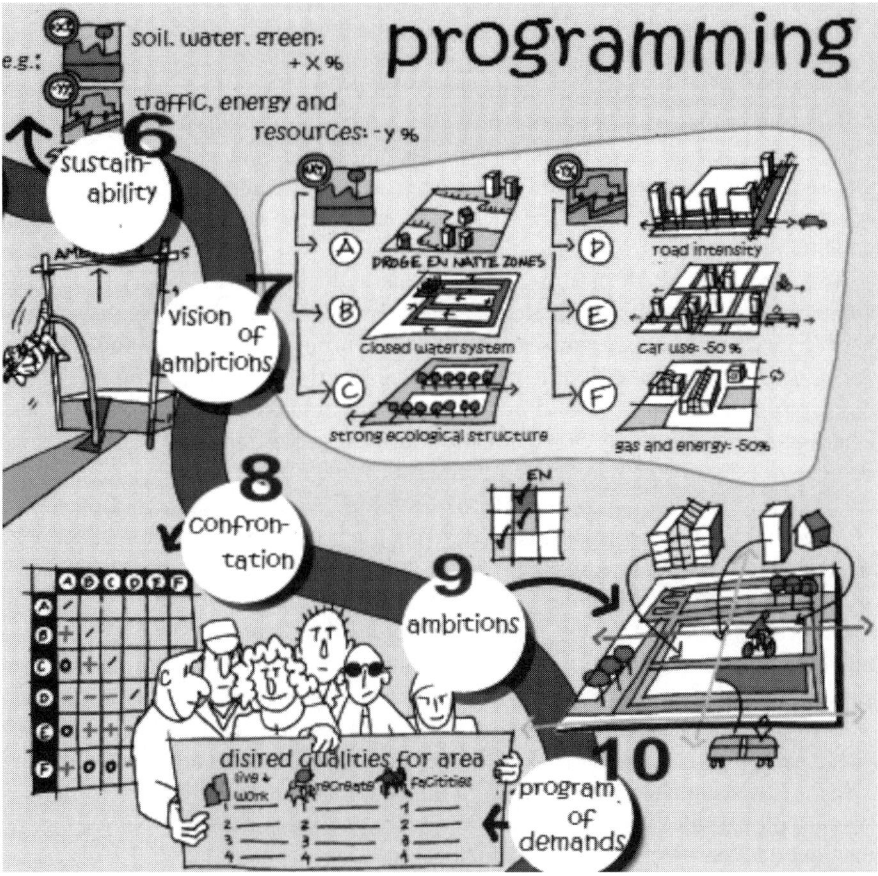

Figure 8.10 The study phase in the MIRUP planning process (Stadsgewest Haaglanden 2003)

Parts of the method are also frequently used outside the Hague region. This is not done by following the MIRUP planning process described above, but as part of the MILO method described in 8.2. Previously, municipal authorities in the Netherlands could obtain support from the VNG's MILO Project Bureau in order to optimize environmental input in spatial planning. The MILO method was produced for this purpose.

MIRUP can also be used to inspire local professionals to use parts of the approach. This may involve the layer approach as an analysis instrument, for example, or the extensive overview of environmental measures. The overview of project examples shows that applying these instruments has indeed resulted in plans whereby environmental policy has made a greater contribution to the quality of the physical environment.

Discussion

The discussion of the MIRUP method has three parts. The first section assesses the position of the MIRUP approach in Dutch environmental policy, based on the framework discussed in Chapter 3. The second part of the discussion involves a more in-depth analysis. Next, the five characteristics of the MIRUP approach will be examined. The final section focuses on MIRUP's scope of application in planning processes.

We have indicated that one of the method's main characteristics is its communicative character. If MIRUP is used in a spatial project, the intention is that the environmental department is involved throughout the planning process, alongside the spatial planning department. During the MIRUP planning process, they work together to find solutions to the spatial issues faced. The environment department can help to prevent a situation in which environmental considerations are obscured by other interests. On the basis of this way of working, the method can be placed in the lower right section of the framework (Figure 8.11a).

The first characteristic to be considered is *intention*. This refers to the purpose of the methodology, as envisaged by its originators. The objective of MIRUP is defined as offering 'handles' to incorporate environmental aspects in spatial plans. The explanation of the method expressly states that the MIRUP is not a blueprint but a guideline. In terms of the characteristic *intention*, the method can be positioned in the outer ring of the model.

MIRUP offers sufficient scope to include all relevant environmental aspects when searching for a solution to an issue. This means that, in addition to themes from the 'grey' environment, themes relating to the 'green' and 'blue' environments can also play a role. In terms of the characteristic *integration*, the method can be positioned in the outer ring of the model.

The MIRUP method must be applied in close consultation between the spatial planning and environmental departments. The departments carry out the analyses jointly, decisions are taken jointly, and the target scenarios should also be a joint product. The method offers a series of building blocks for structuring the cooperation. In terms of the characteristic *interaction*, the method can be positioned in the outer ring of the model.

The MIRUP method is used to develop area-specific solutions for spatial issues. It provides a number of area types that can be used to define environmental ambitions for the plan that is under discussion. Users can also define additional area types or adapt existing ones. In terms of the characteristic *differentiation*, the method can be positioned in the outer ring of the model.

The final characteristic is *conformance and performance*. The ultimate objective is to produce a spatial plan in which environmental interests are properly represented. Relevant environmental aspects even help to shape the solution for the spatial issue. In terms of the characteristic *conformance and performance*, therefore, MIRUP can be positioned in the outer ring of the model. Figure 8.11b depicts the relationships between the characteristics.

The third part of this discussion focuses on the applicability of the approach in the spatial planning process. In this respect, MIRUP is in a somewhat special position – not so much in terms of its actual scope, but in terms of the 'how'. MIRUP provides a structure for the spatial planning process: the MIRUP planning process. The structure comprises all the steps that are normally found in planning processes, from study to implementation. However, MIRUP offers no basis for the final phase. Figure 8.12 gives a total picture of the aspects in this discussion.

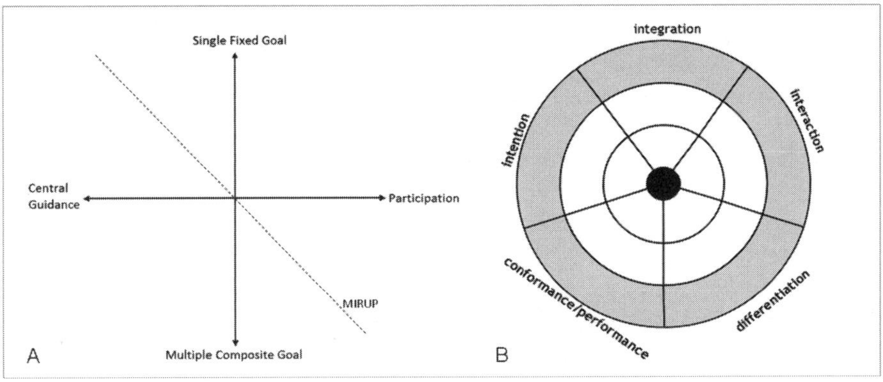

Figure 8.11 Characteristics of MIRUP

Conclusion

The survey of the method reveals various strengths and weaknesses. One of MIRUP's strengths is that it provides a basis for structuring the spatial planning process from beginning to end, without losing sight of environmental considerations. The environment remains on the agenda throughout the process. Another strong point is the imagery that is used (Figure 8.12). This enables actors from diverse contexts to understand each other more easily. This effect is reinforced by the many examples used to illustrate each step of the MIRUP planning process. In addition, there is an extensive overview of possible measures for resolving the spatial issue. Finally, the communicative nature of the instrument means that the method calls for a proactive approach on the part of all those involved. It thus promotes the cohesion between the various policy departments of the local authority.

MIRUP also has a number of weaknesses, however. These are partly inherent in the goal of the method and the choices that were made when it was designed. For example, the method was designed for use by municipal authorities in the Hague region, and area types are defined accordingly. This means that the method cannot be directly applied to other locations in the Netherlands. Another weakness of the method is its somewhat simplistic approach to delivering sustainable futures. After all, the method assumes that incorporating environmental aspects in spatial plans automatically means that these plans are sustainable.

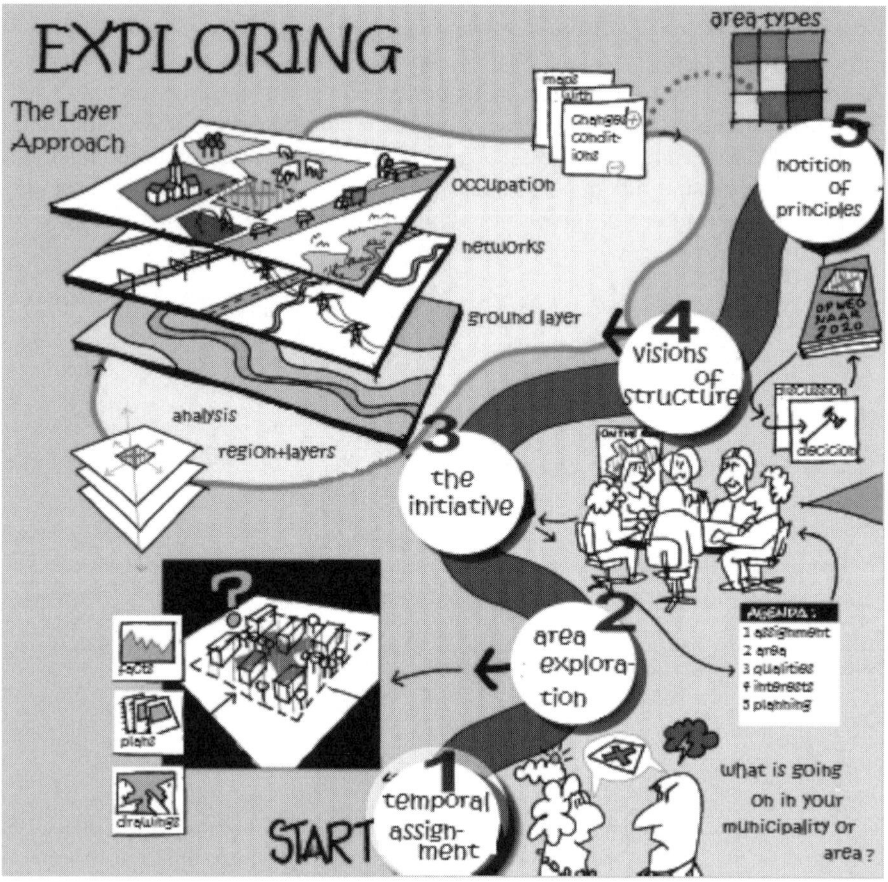

Figure 8.12 Imagery used in MIRUP: communication between actors (Stadsgewest Haaglanden 2003)

Despite its weaknesses, the area-specific and flexible approach of MIRUP mean that it fits in well with decentralized environmental policy. The method is based on area-specific customized solutions to spatial issues.

8.4 Zip Methods in Perspective

Zip methods can be seen as products of decentralized environmental policy. They are not based, for example, on inflexible policy goals[6] that must be realized at any cost. Instead, these methods consider the specific characteristics of areas as well as their potential, the opportunities that exist and the possible threats faced.

6 Obviously, statutory requirements are not ignored.

Area-specific quality ambitions are defined on the basis of this information, where possible in consultation with all stakeholders.

The methods discussed in this chapter all adopt an area-specific and communicative approach. They also share a reliance on imagery that enables actors from diverse backgrounds to communicate easily with each other. This means that they are not only easy to use during communication, but are also suitable as presentation instruments.

When the three methods are compared, various aspects draw our attention. First, there is an important difference between LOGO on the one hand, and MILO and MIRUP on the other hand. MILO and MIRUP provide a structure whereby environmental policy provides input to the spatial planning process; the policy fields are 'zipped together', as it were. This means that the user has the right support instruments for the process at the right time. This is less true of the LOGO method, which focuses on defining quality ambitions. Process-related aspects are confined to an appendix. Another difference is the way in which the methods deal with relevant terminology. Each method is designed to ensure the optimum integration of environmental considerations in spatial planning processes. This is made clear in LOGO and MILO. However, MIRUP is not clear on this point. The terms 'environment' and 'sustainability' are frequently interchanged. This can be confusing.

All the 'zip' methods discussed in this chapter are relatively new methods. They have only existed for a few years. This means that less experience has been gained in using them than in the case of the VNG zoning method, for example. Nevertheless, the practical application of the methods has shown that there are benefits to be gained by 'zipping together' the relevant policy disciplines.

Chapter 9
Information Methods: Showing is Knowing

Dutch environmental policy has changed in many respects in recent years. The national government has formulated proposals for decentralization and deregulation with the idea of creating greater scope for locally oriented solutions. This would imply that local authorities have more options for protecting and enhancing the quality of the physical environment.[1] There are developments in other policy areas too. For example, the frameworks for Dutch environmental policy are increasingly defined in a European context.[2] Guidelines for air quality are one example. Less well known, but no less relevant, is the Aarhus Convention,[3] which stipulates among other things that information on the environment must be made public. These directives have far-reaching consequences for the way in which environmental policy is elaborated in the Netherlands.

Both these developments – decentralization and the increasing influence of Europe – have contributed to what, for the Netherlands, is a new category of methods for environmental externalities: environmental atlases. These are instruments that are designed to gather, structure and present environment-related information. Although they are a relatively new policy instrument in the Netherlands, the first experiences appear to be positive. In the municipality of Deventer, for example, the use of the environmental atlas has helped to bring the departments involved closer together, thereby facilitating greater coherence between spatial planning and environmental policy. In this respect, the environmental atlas fits in with the wish of the Ministry of VROM for a more tailored instrument (2006a; 36), which is necessary to strengthen the coherence between spatial policy and environmental policy. This also applies to the wish to bring environmental policy closer to the citizen (2006a; 48).

---

1 In Chapter 7, Environmental Quality Profiles are characterized as methods that align well with a decentral policy environment. Their application results in area-specific solutions.

2 Recent research has shown that 'Europe has a considerable influence in the area of the environment (i.e. in at least 66% of national regulations)' (Douma et al. 2007; 42). These figures are for the Dutch situation.

3 The Aarhus Convention was drawn up in 1998 by the United Nations Economic Commission for Europe (UNECE). The European Union then drew up a directive (2003/4/EC) to comply with 'Aarhus'. In 2005 the Dutch government, in its turn amended the Environmental Management Act (WM, *Wet Milieubeheer*) and the Government Information (Public Access) Act (WOB, *Wet openbaarheid van bestuur*) in order to comply with the European directive.

Experiences in other countries have shown that environmental atlases can improve the coherence and coordination of spatial policy and environmental policy.[4] The same applies to the dialogue with the local population: an environmental atlas – like certain Quality Profiles (see Chapter 7) – can be used as a communication instrument. However, although other countries have worked with environmental atlases for up to twenty years, the interest has only arisen in the Netherlands in recent years. Section 9.1 discusses one of the very first Dutch environmental atlases, that of Deventer. Section 9.2 focuses on the *Guide to Compiling an Environmental Atlas*. This is not an environmental atlas that can be directly applied, but a development plan for local authorities that wish to develop and use their own environmental atlas. In Section 9.3, the Deventer Environmental Atlas is compared with the suggestions in the *Guide to Compiling an Environmental Atlas*.

9.1 The Deventer Environmental Atlas

The history of the Deventer Environmental Atlas begins in 2002, the year in which the possibilities for an environmental atlas at local level were first explored. After a brief experimental period, it was decided to link the first version to the new Deventer environmental policy plan that was also being formulated at the time. Both were completed in 2003 (Gemeente Deventer 2003a, 2003b). The link with the new environmental policy plan meant that the environmental atlas was anchored in policy from the outset, so that its use was not optional.[5]

At the time, Deventer was the only municipality in the Netherlands that wished to develop an environmental atlas. This wish was mainly the result of long-standing problems regarding the coordination of and coherence[6] between spatial planning and the environment. Too often, the Environment department was consulted too late in spatial planning processes. This occasionally led to difficult situations, particularly if land-use plans did not take sufficient account of the environmental space taken up by industrial activities (interview with Nijssen, 2006a).

A second reason was frustration with the lack of information, communication and coordination. Information about the physical environment is usually spread between several different departments. One particular department is responsible for noise measurements, while others are responsible for identifying polluted soil

4 Visser and Zuidema (2007) have studied three leading environmental atlases of three countries in order to see what can be learned for the development and application of environmental atlases in the Netherlands. The atlases studied were those of Berlin, Prague and Vitoria-Gasteiz (see also Section 10.2).

5 In practice, however, this anchoring is not as strong as it seems. The maps in the first version of the environmental atlas are primarily 'a spatial translation of the environmental themes from the Environmental Vision 2003-2008' (Gemeente Deventer 2–3a; 1). This vision for the environment, like the environmental atlas, is part of the environmental policy plan.

6 See also Chapter 2.

locations or air pollution. A location-selection study requires all this information and more. At the end of 2003, a location-selection study was carried out in Deventer for the Steenbrugge district. The Environmental Atlas had been completed some time before this. Thanks to the atlas, the municipal authority could produce the required information in no time (De Weerd 2004). Before environmental atlases existed, this information had to be collected for every new spatial project. This is usually a time-consuming and expensive task. Finally, the third reason for producing the Environmental Atlas was the municipal authority's wish to be able to inform the local population in a proactive way about the quality of the physical environment[7] (Figure 9.1).

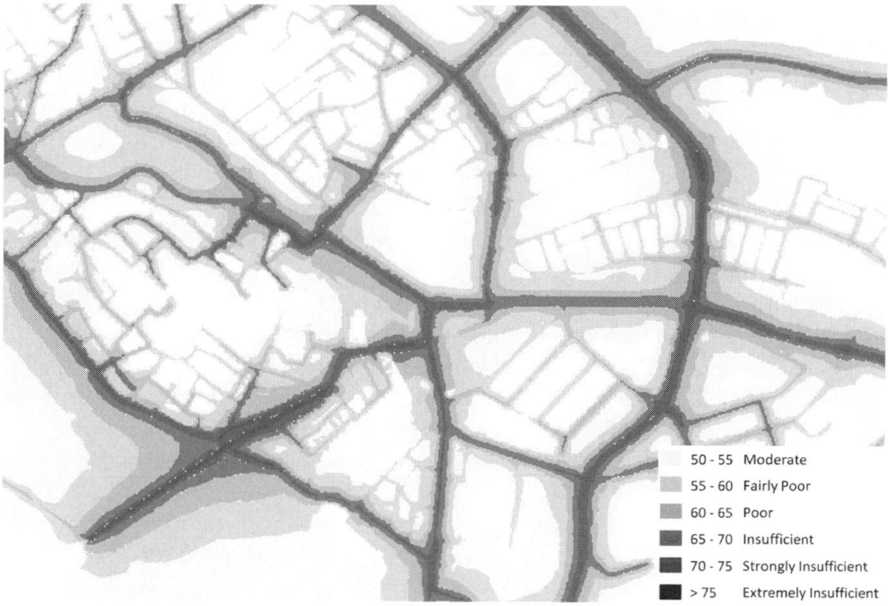

Figure 9.1 Noise levels related to traffic noise in Deventer (Source: Milieuatlas Regio Stedendriehoek, see: http://milieuatlas.regiostedendriehoek.nl)

The Environmental Atlas has now been in use in Deventer for several years. Although its introduction was greeted with a degree of scepticism, the experiences have been positive (interview with Nijssen, 2006a). The instrument is now used consistently in new spatial projects, thereby preventing a situation in which

7 However, this interaction was not fully achieved with the first (printed) version of the atlas (De Weerd 2004). The gap between citizens and government proved too wide. In the Future Agenda for the Environment, this is also used as an argument for initiating the development of digital information systems for health and the environment (VROM 2006a).

environmental aspects are considered at too late a stage in the planning process. The spatial translation of the relevant environmental themes facilitates this: '... we [have] proved able to communicate on the same wavelength with colleagues in spatial planning. And that is a bonus.' (interview with Nijssen, 2006a). At the beginning of 2007, the transition was made to a digital version available on the internet. Local residents are expected to take a greater interest in the digital version of the atlas than in the first printed version.

Procedure

The Deventer Environmental Atlas occupies a special place among the other methods for environmental externalities discussed in this book. The application of environmental zoning methods, for example, can result in a zone within which restrictions are placed on spatial development. Methods in the Quality Profiles category can be used to formulate a vision for an area. These are methods in which certain forms of input produce a particular result. In the case of the Deventer Environmental Atlas, this is less explicit. In the first place, the atlas is – as mentioned above – a combination of environment-related information available within a municipality. In other words, it is not intended to provide answers but simply to inform.

The information is used to support planning and decision-making relating to spatial development. Although the atlas does not therefore produce concrete solutions to planning issues, the instrument contributes to the solution for a spatial task. When the atlas was developed, account was taken of the wish to help steer the resolution of spatial issues from an environmental perspective. The maps in the Environmental Atlas therefore are informing and are to be used as indicators expressing difficulties for spatial initiatives due to environmental spillovers. They are not to be used as conditions setting. A distinction is made between the statuses 'Inform', 'Take into account' and 'Comply' (Gemeente Deventer 2003a).[8]

Maps with the status 'Inform' have a purely informative function. The information in such maps confers no obligations whatsoever. If a map has the status 'Take account of', the information cannot simply be ignored. Deviation is only possible if good reasons are given. For example, in the Environmental Atlas of the municipality of Deventer, the map 'Indicative potential locations for wind energy' has the status 'Take account of'. Claims on these locations for functions other than wind energy will have to be substantiated. The status 'Comply' takes things one step further. This is the most binding status for a map. Information on maps with this status has a legal foundation and is therefore determining in terms of policy formulation. The map '*Beleidslijn ruimte voor de rivier*' ('Space

8 For each map, the decision is made as to the level at which it should be adopted (College of Burgomaster & Aldermen or municipal council). Maps with a proposed non-binding 'Inform' status are submitted to the College for approval. Maps with a more binding status are submitted to the municipal council.

for the River policy guidelines') is one of the maps with the status 'Comply'. In areas to which these policy guidelines apply, new development projects are assessed against the Public Works (Management of Engineering Structures) Act (*Wet beheer rijkswaterstaatwerken*).

The digital Environmental Atlas for Deventer is based on the same principles as the first printed version. The digital version uses the same statuses. It provides extra functions, however. The maps are presented using a GIS viewer (Figure 9.2). It is possible to 'overlay' different map layers – something that was not possible with the printed version. The GIS viewer can also zoom in on specific areas, whereas the first version of the map was a 'static' overview.

Figure 9.2 GIS application for the Deventer Environmental Atlas

In principle, there is no limit to the number of themes that can be included in the Environmental Atlas, although its name indicates that it is designed to provide environment-related information. Initially, other themes are not considered.[9] When the first printed version was published, the Deventer municipal authority indicated that the development of the atlas is based on a growth model, which means that new themes can be added in the future.

The first version of the Deventer Environmental Atlas covers the following themes: soil, noise, odour, dust, air, radiation, water, waste, energy, green spaces

9 It is always possible to include other themes apart from environmental information. These include socio-economic factors that can affect the physical environment. This is discussed in Section 9.2.

in the residential environment, nature and ecology, municipal environmental-audit information, public safety and security, traffic and transport, building and housing and – finally – spatial planning. Some of these themes were still being developed at the time of publication. The growth model is reflected in the new digital version of the atlas, which includes new themes (e.g. operating licences issued). Some of the other themes are divided into sub-themes. The atlas also includes non-environment-related information, such as cadastral information and district maps.

Practice

The atlas has proved to be a success (interview with Nijssen, 2006). Officials of the Deventer municipal authority have been using the atlas since it was published in 2003. Since that time it has served as a source of information to support decision-making and spatial development. We have already mentioned the example of the Steenbrugge district.

However, support for planning and decision-making is not the only function of the Environmental Atlas. As mentioned, it is also a communication instrument. The atlas has helped to improve internal communication between the various departments within the municipal authority. It also has the potential to inform the public more effectively and hence improve communication between the municipality and the public. The first printed version was, however, not very successful in terms of improving communication with residents and stakeholders.[10] The Deventer municipal authority hopes that the new digital Environmental Atlas, which has been available since 2007, will bring policy for the physical environment closer to the citizen. The 'ingredients' are at least in place.

Research has shown that the relevance of environmental policy for the public is strongly determined by the quality of the environment in which they live. The quality of the various relevant themes is expressed in the Environmental Atlas. The expectation is that making the digital atlas available on the internet will bridge the gap between citizens and government. But in this case, too, time will tell. It is in any case easier for residents to obtain environment-related information about their area (see Figure 9.2).

Discussion

The discussion of the Deventer Environmental Atlas has three parts. The first section assesses the position of the approach in Dutch environmental policy, based on the framework discussed in Chapter 3. The second part of the discussion involves a more in-depth analysis in which five characteristics of the Atlas will

---

10 During the period in which the first version was available for perusal at the municipal hall, only a few residents came to inspect it (De Weerd 2004). Informing the public through this atlas proved to be too laborious. Later, too, very little interest was shown (interview with Nijssen, 2006a).

be examined. The third section focuses on the method's scope of application in planning processes.

The diagonal of the framework represents the development of environmental policy in the Netherlands (see Figure 9.4a and Section 3.1). A position in the upper left-hand quadrant is associated, among other things, with the wish to achieve a maximum policy result. The lower-right section represents the aim to create an optimum policy process. The Deventer Environmental Atlas can be placed in this last part of the framework (Figure 9.3a). The instrument is not designed for explicit policymaking, but to support or optimize decision-making in spatial planning.

In the introduction, the Deventer Environmental Atlas is described as an instrument that – in contrast to other methods for environmental externalities – does not result in concrete solutions for spatial issues. This can also be seen in the various characteristics. It is not possible to give an unequivocal answer for the characteristic *intention*. A map with the status 'Comply' will carry more obligations than one with the status 'Inform'. This means that, in terms of the characteristic *intention*, the method can be placed in any ring of the model, depending on which map is being discussed.

The map 'Indicative potential housing locations' combines all the relevant policy themes, thus providing information on areas that are not within an ecological zone or are exposed to some degree of environmental impact. In terms of the characteristic *integration*, the method can therefore be positioned in the outer ring of the model.

A third characteristic is *interaction*. The Deventer Environmental Atlas is designed to be more than a policy-supporting instrument for internal use. Another important function is that of communication instrument in respect of third parties, including citizens and businesses. The municipal authority hopes the atlas will stimulate the dialogue with other sectors. This means that, in terms of the characteristic *interaction*, the method can be placed in the outer ring of the model.

The Deventer Environmental Atlas does not make a direct distinction between area types, unlike the various Quality Profiles. Indirectly, area differentiation is possible, but this is done by the user. The user can distinguish between various areas on the map and see the environmental considerations for each area. An industrial area will be regarded as less sensitive than, for example, areas within the Ecological Main Structure (EHS, *Ecologische Hoofdstructuur*). This can be taken into account during policymaking. In terms of the characteristic *differentiation*, therefore, the Environmental Atlas can be placed in the outer ring.

The status given to a map determines the degree of *conformance and performance*. Maps with the status 'Inform' can be excluded from the decision-making process. This is not possible for maps with the status 'Comply'. This means that, in terms of the characteristic *conformance and performance*, the method can be placed in any ring of the model. Figure 9.3b shows the relationships between the characteristics of the Deventer Environmental Atlas.

It is a policy instrument that can be of added value in all phases of a spatial-planning process. Environmental considerations can be introduced during the preparation of the process, and monitored in later phases. It is conceivable that the Environmental Atlas can be consulted at any time to substantiate arguments, verify claims or simply to check the situation with regard to a particular environmental aspect. Subsequently, the process can be evaluated against the intentions, and the maps can be used to inform local residents about the plans, etc. The way in which the atlas is used depends on the level of detail and how up to date it is. In order to serve as an evaluation instrument, for example, the maps must be regularly updated.

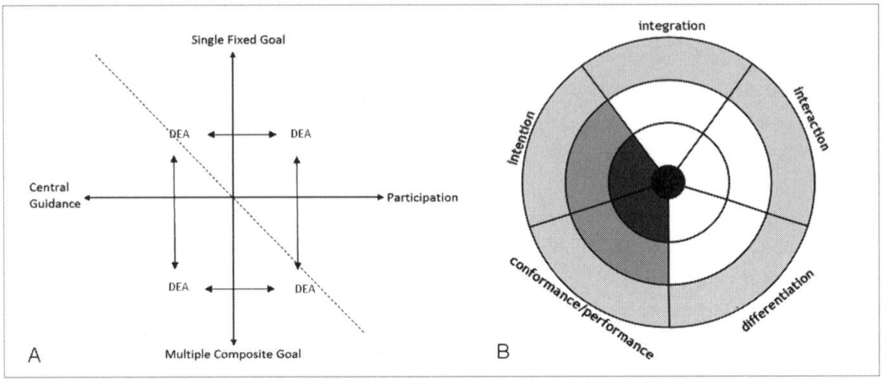

Figure 9.3 Characteristics of the Deventer Environmental Atlas

Conclusion

The survey reveals various strengths and weaknesses in the instrument. One of its strengths is that, once completed, the Environmental Atlas provides fast and accessible information about the quality of the local environment. The original variant has been shelved and the modern digital version is available via the internet or intranets. Another strength is that the Environmental Atlas can improve internal communication within local authorities. The Atlas presents information about environmental aspects (measurements, restrictions, etc.) from a spatial perspective. Environmental officials can therefore communicate in the same languages as their colleagues in spatial planning. This can have a positive influence on the coordination of spatial planning and environmental policy. The Environmental Atlas is also a suitable instrument for external communication: maps with gradations appeal more to the public than tables of figures. Finally, we have mentioned the developments regarding GIS software, which contribute to greater accessibility and allow the user to combine and/or compare data.

In general, an environmental atlas is a relatively new policy instrument in the Netherlands. It is therefore difficult to estimate its value, now and in the long term. One aim of such an atlas is to promote the dialogue between citizens and local authority, but information can be wrongly interpreted, leading to misunderstandings. Information can also be deliberately misinterpreted, resulting in certain issues being 'blown up'. A further aspect to consider is the gathering and processing of relevant information. Most of the information is available within a municipal authority, but is usually spread between departments. In addition, the information has to be processed before it can be incorporated in an environmental atlas. This can be a complicated and time-consuming process.

The initiative of the Deventer municipal authority has also been taken up by other municipalities in the Netherlands. Several municipalities have an environmental atlas or are developing one. There are developments at other levels of government too. In the province of Gelderland, the physical environment has been digitally mapped, and for several years the Interprovincial Council (IPO) has coordinated the provincial risk maps. With an environmental atlas it is also possible to participate in national projects such as DURP[11] or PASTIS.[12] An environmental atlas thus seems to be a logical next step, as part of Dutch environmental policy. However, for some municipal authorities this is too big a step.[13] There is little experience to draw on in order to develop an environmental atlas. The following section discusses guidelines for developing an environmental atlas.

9.2 Guide to Compiling an Environmental Atlas

Unlike the instruments already discussed in this book, the 'Guide to Compiling an Environmental Atlas' is not a method that can be directly applied to obtain concrete solutions to spatial issues. Neither is it directly applicable in the same sense as the Deventer atlas. Instead, this Guide explores the possibilities for using environmental atlases in current policy for the physical environment (Visser and Zuidema 2007).[14] On that basis, and on the basis of experiences with existing environmental atlases in the Netherlands and other countries, a step-by-step plan is presented that local authorities can use to compile their own environmental atlas.

The Guide was based on two developments. The first relates to developments in the policy field relating to the physical environment. The EU Directive 2003/4/EC[15] came into effect on 14 February 2003 and is commonly associated with

11 DURP stands for *Digitale Uitwisseling in Ruimtelijke Processen* (Digital Exchange in Spatial Processes), formerly *Digitale Uitwisseling van Ruimtelijke Plannen.*

12 PASTIS stands for *Postcode als sleutel tot informatiesystemen* (Postcode as the key to information systems). This is a project of the Health and Environment Action Plan and its aim is to provide citizens, on the basis of a postcode, with information on the quality of their physical environment and indoor environment (VROM 2006b; 2).

13 Interview with Nijssen (2006a).

14 The history of the Guide begins in 2001, the year in which the first study was carried out into the feasibility of an atlas on environmental health and hygiene (Van der Linden et al. 2001). The purpose of the atlas on environmental health and hygiene was to provide information on the 'grey' environment.

15 Directive 2003/4/EC is a direct consequence of the Aarhus Convention and contains regulations for access to environmental information (the first 'pillar' of the

the Aarhus Convention on the public availability of environmental information. It stipulates that EU Member States must make environmental information accessible to the public. As a result, the DUIV alliance[16] in the Netherlands launched the Aarhus Implementation Project (PRIMA) to assist central and decentral government authorities in implementing the obligations. The directive has been translated for the Dutch situation in the Environmental Management Act (WM, *Wet Milieubeheer*) and the Government Information (Public Access) Act (WOB, *Wet openbaarheid van bestuur*). As of 14 February 2005, exactly two years after the adoption of the European directive, all government bodies in the Netherlands must comply with the new provisions. However, research has shown that this is not such a simple matter. There is a lack of clarity as to what constitutes 'environmental information' and as to whether the legislation applies to all the information that an organization has (OpdenKamp Adviesgroep 2004; 28).

The second development follows is the result of a changing relationship between citizens and government agencies. Policy is increasingly formulated 'in an interaction between citizens, interest groups, politicians and officials throughout the policy process' (RMO 2003; 37)[17]. Despite this, the VROM Council states that 'a gap has arisen in the communication process between environmental professionals and the citizen' (2005; 52). It is not always easy to interest the public in local environmental policy. As early as 2004, Secretary of State Van Geel wrote to the municipal authorities that 'adequate information provision to citizens is essential in order to assure involvement in environmental policy' (VROM 2004c; 2). Local involvement is an important factor in building support for the policy.

Apart from these two developments there is a need for greater coherence between spatial policy and environmental policy, particularly in situations whereby local authorities have greater responsibility for this coherence as a result of decentralization. Both forms of policy are geared to the quality of the physical environment and both influence each other. Furthermore, there is not always sufficient coordination or consultation. In the Future Agenda for the Environment, the government observes that greater coherence is required 'at the local and regional levels' in particular (VROM 2006; 36). In Chapter 8, MIRUP and LOGO were discussed as instruments that can help to improve this coherence. This also applies – albeit in a different way – to the environmental atlas envisaged with the 'Guide to Compiling an Environmental Atlas'.

Convention). The second pillar of the Convention (participation in decision-making) was implemented in the EU Directive 2003/35/EC.

16 DUIV is an partnership of the Ministry of VROM, the Association of Water Boards (UVW), the Interprovincial Council (IPO) and the Netherlands Association of Municipalities (VNG). See also Chapter 2.

17 In Chapter 3 this is described as a shift towards more communicative policy. In this context, the overhead is working with other parties – to a greater extent than in the past – in order to achieve goals. The various parties together ensure that policy is developed and implemented.

Procedure

An environmental atlas can be produced in different ways. Its definitive form depends to a large extent on the user's requirements. It is therefore crucial to establish the 'ambition level' at an early stage. This should be done in the first step of the step-by-step plan. The Guide defines four ambition levels (see Table 9.1).

At the first level, a 'basic atlas' is compiled. It contains information that municipal authorities often require for plan development. This includes maps with noise measurements, but also Health Effect Screening or IMZ maps. This is what makes the environmental atlas an informative document. The main object of the instrument is to quickly provide an overview of the actual environmental situation in a municipality. Ideally, the atlas also summarizes current legal environmental requirements and is a response to the requirement to make environmental information available. The basic atlas can also provide an initial framework for links to spatially relevant information, thereby establishing a connection with DURP.[18] Working from the basic atlas, it is possible later to choose and develop a new ambition level that is considered appropriate.[19]

Table 9.1 Ambition levels for compiling an environmental atlas (Visser and Zuidema 2007)

Ambition level	Organization/stakeholders
Basic atlas	Preparation and production of the environmental atlas by the designated or initiating government department(s).
Public interpretation	Interpretation, with stakeholders, of the data in the environmental atlas.
Environmental atlas or municipal/ provincial project	Set up the environmental atlas as a 'database' that is accessible to different government departments – use of GIS and municipal/ provincial intranet/network is required.
Environmental atlas as an exchange forum	Set up the environmental atlas as an exchange forum between various government and non-government parties; the atlas is a joint project – use of PPGIS* and an external support bureau would seem to be appropriate.

* PPGIS stands for Public Participation GIS. There is still no clear-cut approach to the use of PPGIS, but its essence is an optimum combination of GIS applications and communication methods that facilitate citizen participation (see for example: Sawicky This is precisely what is envisaged with the environmental atlas as an exchange forum.

18 There is already a digital atlas that presents planned spatial developments: the *Nieuwe Kaart van Nederland* (New Map of the Netherlands). This information system was created through a partnership between the Ministry of VROM and the NIROV (Netherlands Institute for Planning and Housing).

19 The ambition levels are designed to be complementary. This means that it is possible to move from the basic level to more detailed levels.

Choosing an ambition level is only the first step. Subsequent steps can be taken in the development of an environmental atlas (shown in Table 9.2). The steps need not be interpreted in a strict sequence. It can frequently happen that new wishes and requirements arise within the production process. Processes – including the process of compiling an environmental atlas – are generally dynamic and unpredictable. The steps to be followed are designed so that it is always possible to repeat one or more steps to modify the previous design or details. The step-by-step plan that the Guide provides is definitely not intended to be a blueprint. The steps are designed as a guideline and to provide inspiration.

Table 9.2 Possible steps for compiling an environmental atlas

Step	Task	Description
Step 1	*Select ambition level*	Political/organizational definition of the role of the environmental atlas
Step 2	*Organizational structure*	Allocation of responsibilities for initiative, production and maintenance of the environmental atlas, and budget allocation
Step 3	*Baseline review*	Inventory of existing information and comparison thereof with the information required in the environmental atlas
Step 4	*Production*	Gathering and structuring information, preparing it for publication and bundling it in the required format
Step 5	*Publication and distribution*	Provision of and communication about the information gathered for the various fields
Step 6	*Maintenance and updating*	Keeping the environmental atlas up to date or extending it, in direct relation (or not) to an existing issue

It is not certain what the end product will look like. That depends on various factors. In the first place, the chosen ambition level is a determining factor. The way in which the information is collected and processed also shapes the end result. A local authority can also choose to include, or not include, certain information in the environmental atlas. The 'Zeeland environmental register' is, for example, one of the few environmental atlases to contain information about modest pollution (see Figure 9.4). The digital environmental atlas of the ISGO (Goeree-Overflakee Intermunicipal Collaboration Group) is unique in that it includes a link to environmental permits issued to local industry. Information about the environmental health and hygiene aspects of a factory's activities can be displayed on the screen at the click of a button. Spatial translations of complaint registrations are also appropriate for inclusion in an environmental atlas. Another example is the Amsterdam urban district Geuzenveld-Slotermeer, which has used Google

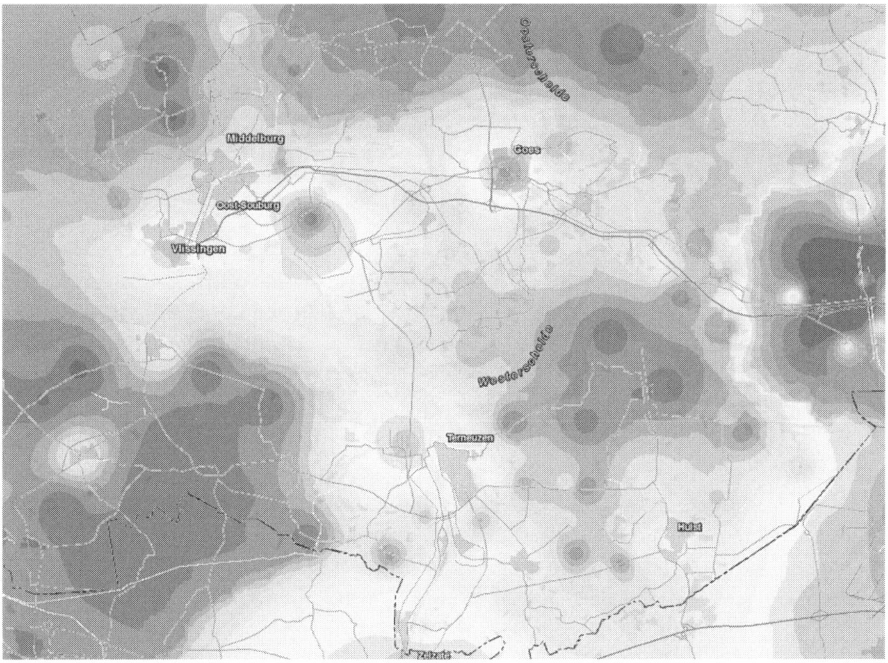

Figure 9.4 Light pollution is one of the environmental themes included in the Zeeland Environmental Register (Provincie Zeeland 2006)

Maps[20] to register environmental complaints and to inform residents (and visitors to the website) about the complaint handling (Figure 9.6). In short, these examples illustrate the flexibility available to decision makers in deciding on the final form of the environmental atlas.[21]

Discussion

The discussion of the Guide has three parts. The first section assesses the position of the method in Dutch environmental policy, based on the framework discussed in Chapter 3. The second part of the discussion involves a more in-depth analysis in which five characteristics will be examined. The third section focuses on the method's scope of application in planning processes.

The diagonal of the framework represents the development of environmental policy in the Netherlands (see Section 3.1 and De Roo 2004). The shift that applies to the Deventer Environmental Atlas is also relevant for the Guide, namely the shift from goal maximization to process optimization. The Guide to an environmental

20 Google Maps is an internet service provided through the Google search engine.

21 The basic atlas that is realized at the lowest ambition level will show little differentiation. The scope for differentiation increases in step with the ambition level.

atlas does not work on the 'input-output' principle: the inclusion of environmental information does not lead to concrete solutions for spatial issues. Instead, its function is to support processes, whereby environmental considerations can be brought into the decision-making. In addition, the Guide clearly reflects the communicative character of the method. On the basis of these characteristics, the method can be placed in the lower right-hand section of the framework (Figure 9.5a).

Information about the defined characteristics (see Section 3.2) can be derived from the position of the environmental atlas in the framework. The first characteristic is *intention*, which refers to the purpose of the method as envisaged by its originators. In the case of the Guide to an environmental atlas this mainly involves – in addition to broadly informing – providing support to spatial-planning processes from the perspective of the environment. This means that, in terms of the characteristic *intention*, the method can be placed in the outer ring of the model.

The characteristic *integration* refers to the degree of coherence between the various environmental aspects and between spatial planning and the environment. This is a characteristic that applies to the Guide to an environmental atlas. Increasing the coherence between spatial planning and the environment is, after all, one of the main aims of the method. In terms of the characteristic *integration*, the method can be positioned in the outer ring of the model.

A third characteristic is *interaction*. This characteristic is also relevant for the Guide to an environmental atlas. Irrespective of the ambition level, the atlas can be used to feed the dialogue with the local population or, for example, to clarify the environmental effects of policy intentions. In terms of the characteristic *interaction*, the method can be positioned in the outer ring of the model.

The fourth characteristic is *differentiation*, which refers to the scope for taking account of differences in area sensitivity. The instrument offers intrinsic possibilities for providing information about area sensitivity. In part, this is inherent to the subject or theme of a map.[22] In terms of the characteristic *differentiation*, the method can be placed in the outer ring of the model.

The final characteristic is *conformance and performance*. This relates to how the use of a method for environmental externalities affects spatial policy. Some methods have a direct effect; with other methods the effect is less direct. The latter applies to the Guide to an environmental atlas. This means that, in terms of the characteristic *conformance and performance*, the method can be placed in the outer ring of the model. Figure 9.5b depicts the relationships between the characteristics.

An environmental atlas can provide added value in the phases of a spatial-planning process that focus on informing. Initially, the instrument can be used to draw attention to environmental interests. It can be introduced whenever necessary in later phases to reflect on choices and intended interventions.

22 An industrial area will be regarded as less sensitive than, for example, areas within the Ecological Main Structure (EHS, *Ecologische Hoofdstructuur*).

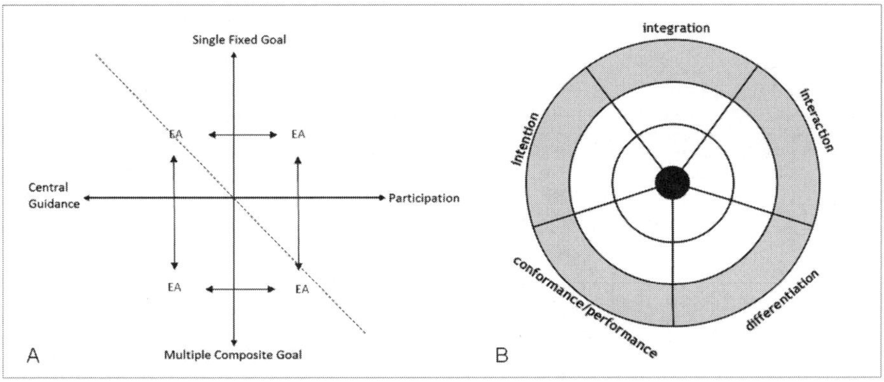

Figure 9.5 Characteristics of the environmental atlas

Conclusion

The survey reveals various strengths and weaknesses in the instrument. The first of these relates to the Guide and not to the intended product. The 'Guide to Compiling an Environmental Atlas' is helpful in providing a relatively uniform and structured approach. Rather than reinventing the wheel in each locality, it thus provides a benchmark that can be used throughout the country.[23] Another point is that environmental considerations can literally be 'put on the map'. Maps are part of the language of spatial planners. Presenting the environment in the same language can significantly improve internal communication within government authorities. The environmental atlas can also provide support for policy monitoring, City & Environment processes and Strategic Environmental Assessment (SEA). Also of interest is the fact that the possibilities offered by GIS software have increased considerably in recent years. More and more packages are being launched in the market, thereby increasing competition and pushing down costs. This means that compiling a digital environmental atlas need not put excessive pressure on budgets. Finally, there are the statutory obligations for public-sector bodies to proactively publish information on the environment. An environmental atlas can help in fulfilling those obligations.

A weakness of the *Guide to Compiling an Environmental Atlas* is that it is still a theoretical instrument. No practical experience has been gained with it, which means that no observations can be made as to its added value. A further point relates

23 Provincial risk maps have been centrally coordinated since 2006. The Ministry of the Interior and the Ministry of VROM have final responsibility for this. The municipal and provincial authorities feed data into the information system under the direction of the RIVM (National Institute for Public Health and the Environment). A national standard, the *Register of High-Risk Situations involving Dangerous Substances* (RRGS) is used to manage the system and simplify information exchange.

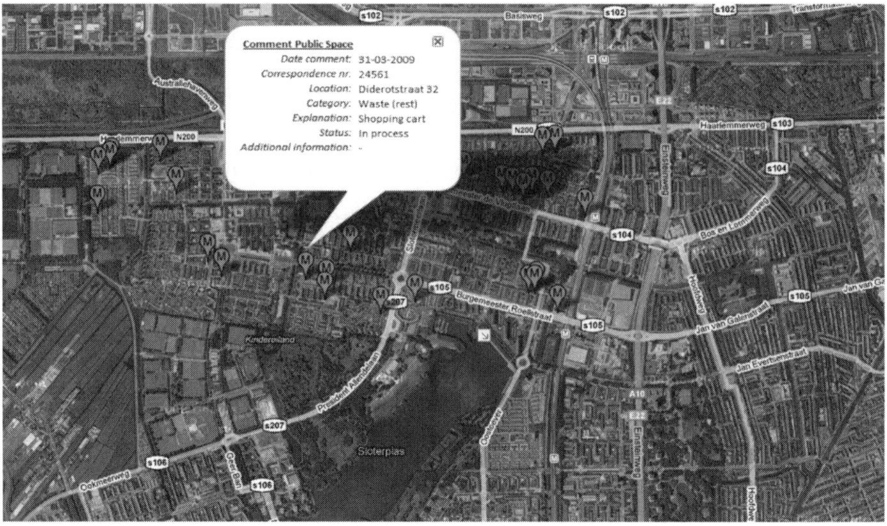

Figure 9.6 Spatial translation of complaint registrations in Geuzenveld-Slotermeer (Source: Google Earth)

to producing an environmental atlas, which, despite technological advances, can still be costly and labour-intensive. Decisions have to be made as to how it will be structured and how it will be embedded in policy, and – not unimportantly – an inventory must be made of information available. Finally, we would point to the need to keep the information in the atlas up to date. This means that, once it has been produced, an environmental atlas will require continuous investment.

The *Guide to Compiling an Environmental Atlas* is a response to the increasing interest in the use of environmental atlases in policy for the physical environment. As a result of various developments, demand seems to be increasing for this type of instrument, which brings together the fragmented information on the physical environment. We have mentioned the obligations resulting from European legislation, the need for greater coherence between the policy fields of spatial planning and the environment, the need for instruments that facilitate information exchange, and the increased public interest in many aspects of the physical environment. Apart from this, the application of other methods for environmental externalities can also be supported. LOGO and MIRUP are examples of methods that require a great deal of information. An environmental atlas can provide support in that context.[24]

24 In a project in 2006, LOGO was used in combination with an environmental atlas (DCMR 2006). This was ultimately an ad hoc environmental atlas, which contained only information on the LOGO project in question.

9.3 Information Methods in Perspective

Shifts in environmental policy over the past decade have contributed to a growing need for information about the physical environment at the local and regional level. For one thing, decentralization and deregulation have led not only to greater policy freedom for local authorities but also to increased responsibilities. Hence it is now the task of municipal authorities to monitor the quality of the physical environment. Second, EU obligations relating to the public availability of environmental information are also an important push factor for the collection and dissemination of this information. Third, and finally, issues are also becoming increasingly interwoven. This can mean that a large number of environmental aspects are relevant for a particular issue. The environmental atlases discussed in this chapter can be of use in providing the required information. Practice shows that, gradually, they are also being used in more and more projects. However, compared to other methods for environmental externalities, environmental atlases are still relatively unknown in the Netherlands. Consequently, little is known about possible policy obstacles.[25] On the basis of practical experience, however, the future prospects for this method appear to be positive.

In this chapter, environmental atlases are presented as instruments for facilitating improved communication and interaction between the policy fields of spatial planning and the environment. Both policy fields are oriented to realizing, improving and maintaining the quality of the physical environment. Interaction between these policy fields at the local level is therefore essential. This begins with knowing what is going on in the relevant field. In this respect, the environmental atlas has already proved its worth for the Deventer municipal authority. The environment has been prioritized more effectively within the process, and internal communication at the local authority has improved. Coherence between policy fields can therefore also be achieved more easily by using a common language. In principle, there is not a great deal of difference between the Deventer Environmental Atlas and the atlas envisaged in the Guide. The Deventer atlas differs in terms of the status accorded to the maps; each status carries obligations that must be met when the atlas is used. Ultimately, this provides certainty for environmental considerations in spatial-planning processes. This is not true, in a direct sense, for the atlas envisaged in the Guide.

The environmental atlas brings some of the traditional environmental assessment methods (e.g. Integrated Environmental Zoning) back into the picture. Chapter 5 emphasizes the possibility of VS-IMZ in terms of providing information about environmental health and hygiene impact in the relevant area. The principles

25 In Chapter 6, the Environmental Performance System was discussed. Initially, this appeared to be an ideal method for linking environmental aspects to spatial planning (in this case, the municipal zoning plan). However, the Dutch Council of State declared this invalid, and the Environmental Performance System is no longer used. The discussion has since reopened (VROM 2006a).

of this method are very appropriate for enhancing an environmental atlas. Maps such as those shown in Figures 5.4 and 5.5 (cumulative environmental impact in existing and new situations) can be usefully included in an environmental atlas. It is even the case that, if the planning consequences of VS-IMZ are set aside, an interesting environmental atlas can be produced. This also applies to the results of certain modern methods for environmental externalities. A map generated on the basis of Health Impact Screening (see for example: Figure 6.4) would certainly not be out of place in an environmental atlas. Similarly, the Environmental Quality Profiles mentioned above usually result in one or more visions for the environment. An environmental atlas can be used to communicate these visions, for example to local residents.

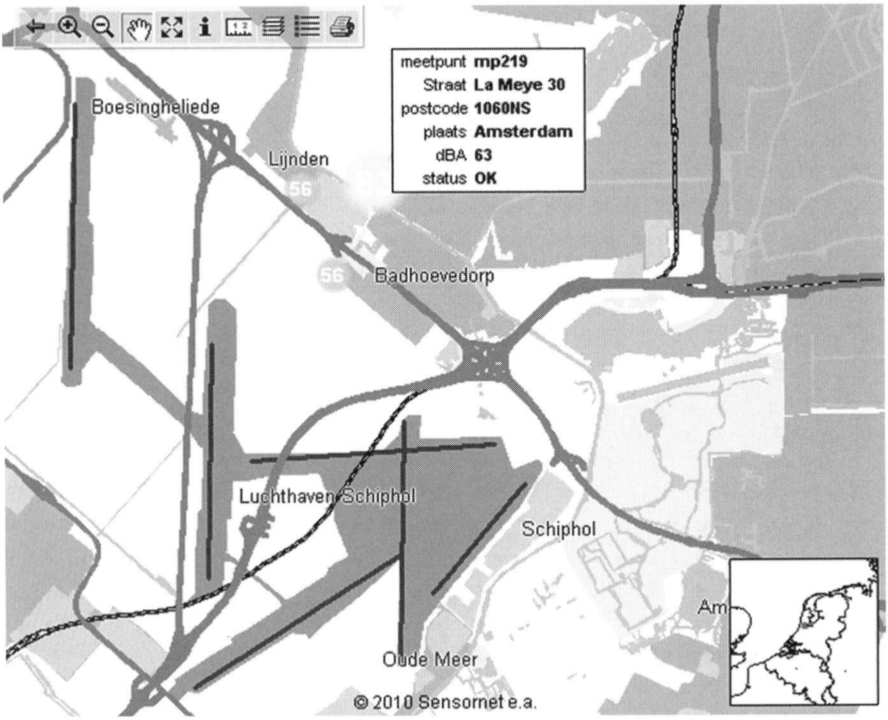

Figure 9.7 Residents want to know about the local situation: current noise measurements around Schiphol

Considering the developments in policy for the physical environment, the environmental atlas would appear to be a logical next step in Dutch environmental policy. As a supplement to existing methods for environmental externalities, it can also help to achieve coherence between spatial policy and environmental policy. However, it is too early for certainties. The environmental atlas is the most recent of the methods for environmental externalities discussed in this book. The instrument will have to prove its worth in the years to come.

Nevertheless, there is cause for optimism. In cities such as Berlin and Prague, environmental atlases have been used successfully for more than fifteen years. In the Netherlands, too, the decentralization process has been accompanied by a number of very interesting initiatives (also Figure 9.7). This type of development is what the Ministry of VROM is aiming for with its Health and Environment Action Plan. Time will tell whether it will also prove to be feasible and of added value.

Chapter 10
Other Methods: Outside the Box

In Chapter 4, the selected methods for environmental externalities were categorized according to several features they have in common. The categories were: Environmental Zoning methods, Checklist methods, Quality Perspectives, 'Zip' methods and Information methods. In addition to these categories there is an 'Other' category for the methods that differ from other methods in terms of their background, approach or procedure. In this chapter we discuss the Environmental Maximization Method (*Milieu Maximalisatie Methode*) (Section 10.1), the ROMBO tactical tool (Section 10.2) and the Bubble Method (*Stolpmethode*) (Section 10.3). A conclusion follows in Section 10.4.

10.1 Environmental Maximization Method

The end of the 1980s saw the widespread adoption of sustainability as a policy concept. Following this adoption, increased attention was paid to the role of long-term effects and hence to coherence between the social, economic and physical aspects of Dutch policies. In only a few years, the use of 'sustainability', and concepts based on it, rapidly became widespread. Sustainability is therefore now seen as a guiding principle when formulating environmental policy plans. The concept has also played a role in the development of various methods for environmental externalities, including the Environmental Maximization Method discussed in this section.

The Environmental Maximization Method is an urban design method whereby environmental themes are used to structure the initial phases of a planning process (Van den Berge et al. 1998). This relates to themes that are in some way relevant from the perspective of sustainable spatial development. The themes used in the method derive from the VNG publication *Bouwstenen voor een duurzame stedenbouw* (Stofberg and Van Hal 1996). The themes are flora and fauna, landscape and soil, water, traffic, energy, household waste, residential environment and, finally, materials used.

The explanation/notes to the method state that 'the idea is that environmental aspects can form an effective starting point for design' (Van den Berge et al. 1998; 9). Hence the defined environmental aspects help to determine the structure of new urban districts. This is an interesting starting point if the aim is to increase the environmental quality of the area. However, in practice there is often relatively little interest in steering aspects based on environmental preconditions and/or conditions. In other words, a method such as the Environmental Maximization

Method can be interesting, but must be accompanied by a strong willingness on the part of stakeholders to use it in comparison to non-environmental priorities.

Procedure

The Environmental Maximization Method has four phases (Van den Berge et al. 1998). The first phase is an inventory or analysis phase that examines relevant location-related aspects and the schedule of requirements. The second phase focuses on how the individual environmental themes can best be incorporated in the spatial design process. In the third and fourth phases, the environmental themes and any preconditions from local authority policy are brought together to form a framework for the eventual urban plan. Figure 10.1 is a schematic representation of the Environmental Maximization Method.

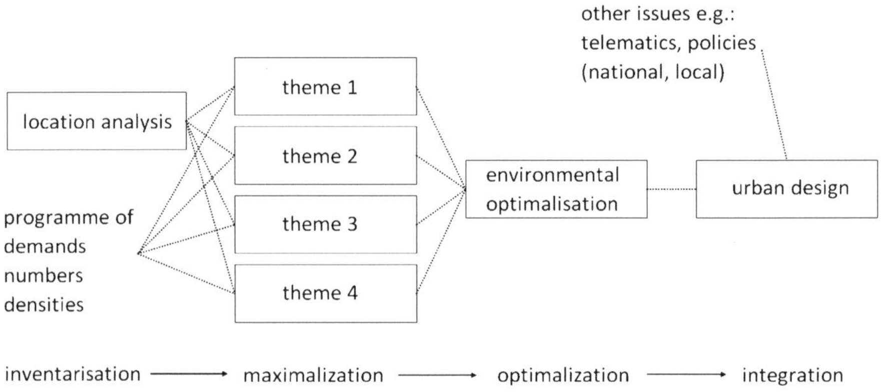

Figure 10.1 Schematic representation of the Environmental Maximization Method (based on Van den Berge et al. 1998)

During the *inventory phase*, information on the current situation in the plan area is collected for each environmental theme. Landscape structure, existing developments and green structures are looked at for this purpose. For the theme 'Energy', for example, the possibilities for combined heat and power systems and collective heat sources are examined. For the 'Materials Used' theme, the Schedule of Requirements for the urban design has been analysed. The aim of this first phase is to identify the possibilities for development and possible problems in doing so.

On the basis of the analysis in the inventory phase, the relevant themes can be separated from the less relevant themes. In the second phase, the *maximization phase*, the first step is therefore to determine the environmental themes with which the design process will proceed; i.e. the themes that have been prioritized. Then, for each theme, it is decided how each theme can be fully incorporated in the final urban plan. This is known as the 'maximization study', and the result is presented in 'maximization maps'.

Each chosen theme results in a map in the maximization phase. In the optimization phase, the separate images are then combined to form a single integrated picture. The maps are placed on top of each other, as it were (Figure 10.2). This process clarifies how the structures and elements of the various themes fit together in a spatial sense. In this phase it may become apparent that the different maximization studies conflict with each other. For example, the plan to build south-facing houses may conflict with the wish to take account of existing landscape structures. In such cases, choices have to be made. These choices are known in the method as 'environmental optimizations'. The result of this phase therefore depends on which parties are involved. In theory, various optimizations can be carried out, for example from the perspective of spatial planning as well as the environment. However, the final choice depends on the opinions and ideas of the stakeholders. Discussion is therefore regarded as an important condition for ensuring that the method is successful.

The final phase is the *integration phase*. In certain respects, this phase can be compared to the optimization phase, which involves relating the chosen themes to each other. In the integration phase, the data on the optimization map are integrated with aspects that are not directly related to the environment. Examples of such aspects are limiting conditions relating to town planning, and requirements set down in municipal policy. This final phase also results in a map image, which serves as the starting point for the final urban-development plan. After this point, the Environmental Maximization Method provides no further input.

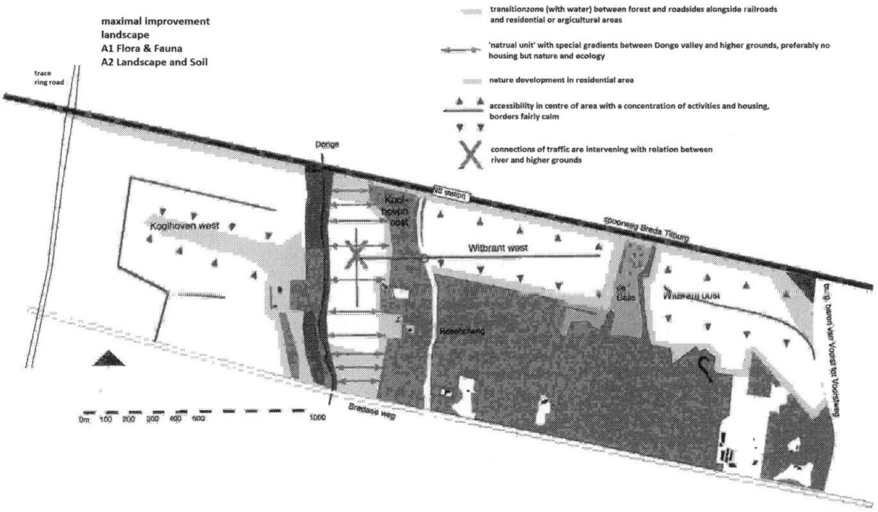

Figure 10.2 An environmental optimization for DE Wijk, Tilburg (Van den Berge et al. 1998)

Practice

The Environmental Maximisation Method has been used in several projects in the Netherlands, one of which served as a trial project for the method: DE Wijk in Tilburg (Van den Berge et al. 1998). DE Wijk is part of the VINEX location of Reeshof, where a total of 12,500 homes are being built. A total of 2,650 of these homes are planned for DE Wijk. The decision to use the Environmental Maximization Method was taken for several reasons. In the first place, sustainability had to be a basic principle, both for the urban development plan and the individual homes. In the second place, the approach chosen at the beginning of the project was one based on providing information rather than setting requirements. The Environmental Maximization Method is in line with this approach.

Maximization maps were produced for four of the method's eight themes: Ecology, Water, Energy and Traffic. The other themes were excluded. In the optimization phase, the four maps were then combined to form an optimum spatial structure for the plan area (see for example Figure 10.2). This map image was the basis for the urban development plan for DE Wijk.

The application of the method in the DE Wijk project is regarded as successful for several reasons. The first reason has already been mentioned: many aspects in the final urban-development plan for DE Wijk correspond to the 'maximum' and 'optimum' structure obtained using the method. Another success factor mentioned is the fact that there was cooperation from the outset between the various disciplines (Van den Berge et al. 1998).

Following the pilot project, the method was used in several other projects, including the restructuring of the Poptahof district in Delft (Van Eijk et al. 2000). In the *Kennisboek Milieu in stedelijke vernieuwing* (VROM 2002a), Dordrecht, Haarlem and Hellevoetsluis are also mentioned as locations where the method was put into practice.

Discussion

The discussion of the Environmental Maximization Method has three parts. The first section assesses the position of the method in Dutch environmental policy, based on the framework discussed in Chapter 3. The second part of the discussion involves a more in-depth analysis in which five characteristics will be examined. The third section focuses on the method's scope of application in planning processes.

A notable feature of the method is its interdisciplinary approach. The various disciplines representing the chosen themes have to work together from the outset in order to achieve results. To this end, no prior requirements should be set regarding the outcome of the process. Instead, as much information as possible is gathered in order to clarify the tasks of the various sides. On the basis of these characteristics, the method can be placed below the centre of the framework (Figure 10.3a).

Now that we have described the position of the method in relation to Dutch environmental policy, we will consider a number of features in more detail. The first characteristic is *intention*. The aim of the method is for parties to work together to find a solution for the spatial task within the specific possibilities of the relevant area, without setting prior requirements from an environmental perspective. In terms of the characteristic *intention*, the method can be positioned in the outer ring of the model.

The method is structured around eight themes that should be representative of the sustainable development of the physical environment. However, these eight themes do not include themes relating to the 'grey' environment, which are also relevant for the quality of the physical environment. In terms of the characteristic *integration*, the method can therefore be positioned in the middle ring of the model.

In the discussion above we have referred several times to the communicative nature of the method. Interaction between different policy disciplines is essential to its success. As mentioned, it is the communication between policy disciplines that is important, as opposed to communication with other actors. In terms of the characteristic *interaction*, the method can be positioned in the middle ring of the model.

The method does not provide direct scope for area differentiation. However, the location features of the plan area play an important role in the method. Therefore, in terms of the characteristic *differentiation*, the method can still be positioned in the outer ring of the model.

The final characteristic is *conformance and performance*. Application of the method results in an outline structure plan that incorporates the possibilities for the various themes. This outline plan serves as a source of inspiration for the urban development plan. In this way, the environmental themes are reflected in the final spatial plan. However, it is strongly emphasized that the method is designed to be a source of inspiration. This means that the results can also be ignored. In terms of the characteristic *conformance and performance*, the method can be positioned in the outer ring of the model. Figure 10.3b depicts the relationships between the characteristics.

The Environmental Maximization Method is designed to be used at an early stage in the planning process. The analytical nature of the method means that it is most useful in the initial phase, in which the possibilities for the plan area are identified. Once the outline structure plan has been produced, the method provides no further input.

Conclusion

The survey reveals various strengths and weaknesses. The strength of the method lies in the fact that the various policy disciplines are 'forced' at an early stage to consult with each other. All the parties involved are therefore encouraged to think proactively about the possibilities for development. Another strength is that the method is not at all dogmatic. There are no limiting requirements, other than those specified in the schedule of requirements for the urban development plan. This is favourable for the creativity needed in planning processes. The method

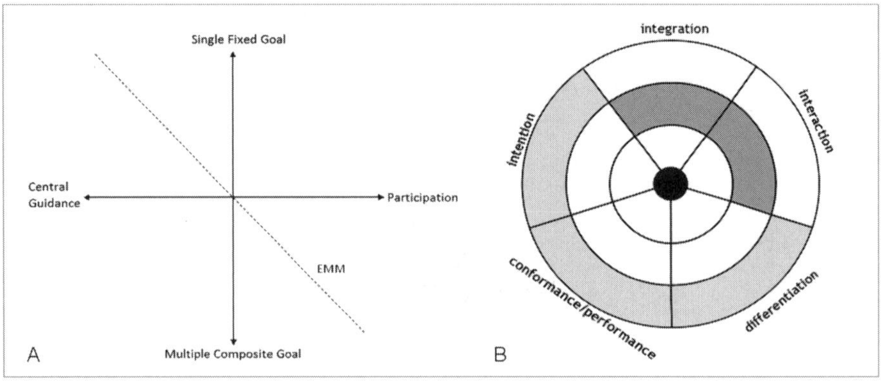

Figure 10.3 Characteristics of the Environmental Maximization Method

is also transparent and well-ordered. Each phase of the process concludes with a map image showing the options. The method is easy to follow for those who are not directly involved. In addition, it can be used as a communication instrument.

The Environmental Maximization Method also has a number of weaknesses, however. It is a time-consuming exercise to make an inventory to produce the outline structure plan. Analysing the various themes and generating map images with solution directions can be a labour-intensive process. Another weakness is that several environmental themes that are relevant for the physical environment do not necessarily need to be included in the process. For example, noise nuisance and air pollution from sources other than traffic are not considered.

Despite these weaknesses, the Environmental Maximization Method is an interesting method for integrating environmental aspects in solutions for spatial issues. The communication-oriented approach of the method means that it aligns well with decentralized environmental policy. The stakeholders from various policy disciplines are encouraged to think proactively about solutions and to contribute to the spatial design process.

10.2 The ROMBO Tactical Tool

The Environmental Maximization Method discussed in the previous section can be regarded as a design method for integrating sustainability aspects in spatial plans. This section focuses on the ROMBO tactical tool,[1] a method that is also intended to facilitate sustainable spatial planning. The two methods differ in their approach, however. Even more central to the ROMBO method than to the Environmental

---

1 ROMBO stands for *Ruimtelijke Ordening en Milieu Beleidsontwerp* (Spatial Planning and Environmental Policy Design).

Maximization Method is the focus on communication and organization in planning processes.

During the 1990s, interest grew in sustainable development and sustainable construction. At that time, however, there were virtually no instruments available to facilitate this. This prompted the municipal authority of The Hague[2] to investigate how open planning processes can contribute to sustainable spatial development (Ravesloot 2002). This led on to the development, between 1999 and 2004, of the ROMBO tactical tool, which was then tested in practice.

The experiment involved examining to what extent the method was able to create conditions for the large-scale realization of energy-neutral[3] building (Ravesloot 2005). Clearly, then, the method was developed with a very specific purpose in mind, i.e. to create conditions for energy-neutral construction projects. An interesting question now is whether the method can be used only for such situations, or whether there are other possible applications. In order to answer this question it is necessary to look at the method's aims in terms of processes. One of the method's main characteristics is that it attempts to bring all the stakeholders to the table, so that they can work together to define and solve the problem(s). This approach means that the method is also relevant for issues in which sustainability and/or energy have a less important role.

Procedure

The ROMBO concept is based on sociocratic design principles.[4] Sociocracy is a system of governance in which all individuals are presumed equal. This is incorporated in ROMBO through the principle that all stakeholders are involved in the process and have an equal say. This means that the stakeholders must be identified before the process begins. In general, the stakeholders are civil servants, architects, project developers and residents. They participate together in a series of workshops in order to establish what they understand by 'sustainability'. The workshops are clustered in three phases: perception, opinion-forming and decision-making (Ravesloot 2005).

The *perception* phase focuses on brainstorming to identify the main possibilities of the area, while also adressing the technical feasibility of ideas and expectations. This is done in three linked workshops. In the first workshop, the project is introduced and the stakeholders can express their perspectives and expectations

2 The municipality of The Hague has undertaken sustainable construction projects since the beginning of the 1990s. In the period 1990-2000, more than thirty special projects were realized (Bakker et al. 1999). Another fifteen similar projects followed in the next four years (Bakker et al. 2004).

3 In the experiment, energy-neutral building was seen as one of the factors that contribute to sustainable urban development.

4 The term 'sociocracy' was thought up in 1945 by Kees Broeke (see also Endenburg 2001).

with regard to the project. The second workshop focuses on conditions for sustainability and new resources for sustainable development. Possible solutions are considered in a brainstorming-type session. During the third workshop, which concludes the first phase, Proposals are made for the continuation of the project and adopted on the basis of 'consent'.[5]

The second phase is one of *opinion-forming*. The aspects considered now focus more on the economic feasibility of the project. This phase also consists of three linked workshops. In the first workshop, alternatives are identified. In the second workshop, the economic feasibility of the alternatives is determined. In the third workshop, the alternatives are ranked in order of preference, with economic feasibility being the main consideration.

Decision-making, the third phase of the ROMBO process, focuses on social feasibility and eventually results in the choosing of an alternative. Again, this is done in three linked workshops. The aim of the first workshop is to provide any necessary clarifications relating to the alternatives and to their economic feasibility. In the second workshop, the alternatives are considered and then ranked on the basis of their pros and cons. In the third workshop, a decision proposal is made, then adopted on the basis of consent.

The proposal adopted in the final workshop is the basis for the further spatial-planning process. ROMBO can therefore be regarded as an instrument for bringing those involved into conformity with the principles, in this case for the purpose of sustainable development.

Practice

ROMBO was developed by the municipal authority of The Hague between 1999 and 2004. This means that the method was created in the policy setting of municipal practice. After the completion of the design process, the method was assessed in order to establish to what extent it was able to fulfil the original objective, i.e. to contribute to the realization of energy-neutral housing.

The pilot project in The Hague showed that it is possible for the government to use ROMBO in order to realize sustainable urban developments. However, applying the method in practice turned out to be a far from simple exercise. Several disadvantages also emerged. In particular, the level of the communicative skills of the workshop leaders proved to be a critical success factor (Revesloot 2005).

Discussion

The discussion of the ROMBO tactical tool has three parts. The first section assesses the position of the method in Dutch environmental policy, based on the

5 Not to be confused with 'consensus', whereby every participant must be in favour of the proposal. 'Consent' means that individual participants 'have no substantiated overwhelming objection to a proposal' (Ravesloot 2005; 178).

framework discussed in Chapter 3. The second part of the discussion involves a more in-depth analysis in which five characteristics will be examined. The third section focuses on the method's scope of application in planning processes.

The ROMBO tactical tool can be distinguished from other methods for environmental externalities on the basis of its almost extreme communicative and intersubjective orientation. The aim is to involve all stakeholders in the process. The goals and principles are formulated and adjusted in the course of the process. On the basis of these characteristics, ROMBO can be placed in the lower right-hand section of the framework (Figure 10.4a).

To a certain extent, the extreme character of the method is also evident in the five characteristics. ROMBO is not a method that is based on a framework of standards for realizing policy goals. In terms of the characteristic *intention*, the method can be positioned in the outer ring of the model.

The aim of the method is sustainable urban development. This means that relevant environmental aspects must be incorporated in the spatial planning process. In terms of the characteristic *integration*, the method can be positioned in the outer ring of the model.

A third characteristic is *interaction*. The communicative nature of the method has been referred to several times: all stakeholders can contribute to the process. In terms of the characteristic *interaction*, the method can be placed in the outer ring.

The characteristic *differentiation* is hardly or not at all relevant for ROMBO and is therefore not included in the framework.

The final characteristic is *conformance and performance*. Use of the ROMBO method is intended to lead to sustainable urban development. The result of the ROMBO process therefore serves as the basis for the further spatial-planning process. In terms of the characteristic *conformance and performance*, the method can be positioned in the outer ring of the model. Figure 10.4b depicts the relationships between the characteristics.

ROMBO is described as a method for supporting sustainable urban development. The process leads to a number of principles for the plans. The ROMBO process must therefore be initiated at an early stage, preferably before the actual design process begins.

Conclusion

The discussion reveals various strengths and weaknesses. A strength of the method is that all stakeholders can contribute to the solution for the spatial task. Not only 'powerful' actors such as project developers, housing corporations or investment companies, but also organized neighbourhood groups can put their stamp on the process.

A weakness of the method is that its success depends to a large extent on the abilities of the workshop leaders. If they are not able to lead the workshops according to the rules of the method, the outcomes may be different – and possibly less appropriate. Another point is that ROMBO assumes that decision-making is

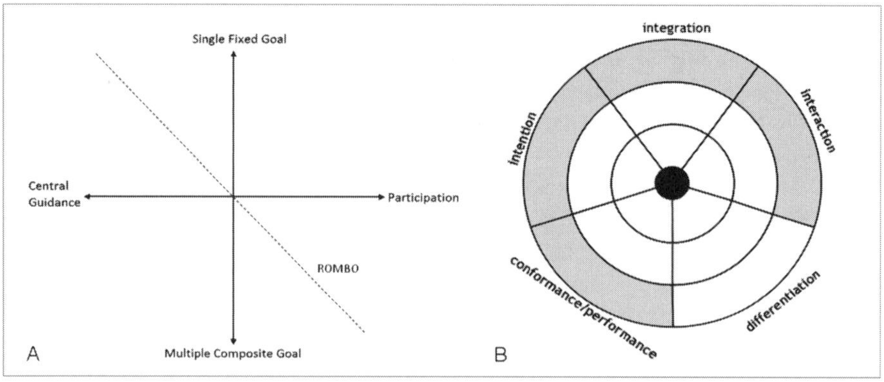

Figure 10.4 Characteristics of the ROMBO tactical tool

based on the principle of equality among the participants. In practice, however, this will not always be the case. Government bodies in particular have difficulty with the idea of entering into a process that has no set goals, whereby the process itself is more important than an envisaged result. The final weakness of the method is that not everyone believes in its effectiveness. This means that there may be only limited support for it, or none at all.

The ROMBO concept is an interesting one. Particularly in the case of complex issues, for which fixed procedures are less effective, ROMBO can contribute to solutions. By now it will be clear that the original aim of the ROMBO tactical tool – i.e. to realize sustainable urban development – does not preclude its use for other issues.

10.3 Amsterdam's 'Bubble Method'

In the early 1990s, many urban development projects were designed according to the 'compact city' concept, the aim of which was to realize a compact, multifunctional and sustainable spatial environment in cities (Bartelds and De Roo 2005). The idea behind the concept is that, by combining and mixing urban functions, the pressure on surrounding rural areas can be eased. The other side of the coin, however, is that functions that are not easily compatible will be located close to each other. This can lead to problems with the quality of the urban environment. The 'Bubble Method' (*Stolpmethode*) was presented in 1993 as a method for weighing up possible solutions to these problems (Rosdorff et al. 1993).

The method was developed for the Environmental Service of the Amsterdam municipal authority. The Bubble Method is introduced in the 1994 Policy Document on Spatial Planning and the Environment (*Beleidsnota Ruimtelijke Ordening en Milieu*) as a method for clarifying 'possible spatial and environmental considerations (…) so that choices can be made that yield the greatest benefit for the environment and the economy' (Gemeente Amsterdam 1994a;9).

The idea of a 'bubble method' is not new. The 'Bubble Concept' originated in the United States, where it is used to establish land-use boundaries for industrial zones.

Procedure

The method is used in the Netherlands to reduce, as effectively as possible, the total level of pollution in a given area. Substitution and compensation of environmental loads are instruments that can be used to this end. The method was refined for the categories 'Noise', 'Risk' and 'Air Pollution' in a case study (Gemeente Amsterdam 1994a).

In order to reduce total environmental load in an area, it is necessary to establish the level and composition of the load in the area. The method consists of several steps for obtaining this information. First, it identifies categories of environmental pollution. The sources of each type of pollution are identified, then the amount of pollution caused by each source.

Once the environmental loads have been established for each category, they are indexed and then combined. The height of the urban 'bubble' is determined by multiplying the combined (indexed) environmental loads by the number of persons affected. The value obtained from this calculation can be placed as a figurative 'bubble' (*stolp*) over the whole city (Figure 10.5). The height of the bubble serves as a reference value. In a similar way, it is possible to indicate the health limit and target value.

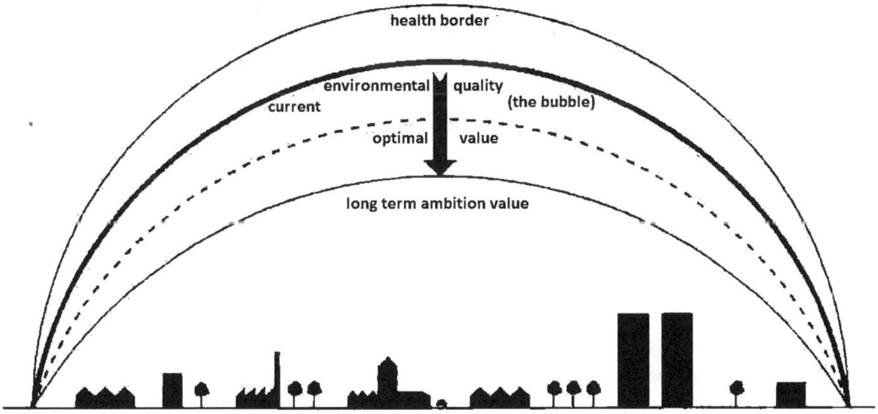

Figure 10.5 Schematic representation of the total city 'bubble' (Gemeente Amsterdam 1994a)

Once the height of the bubble has been determined, the next step is to investigate measures for reducing the environmental load. This can be done generally in two ways. First, by reducing the index. This method achieves an actual decrease in the total environmental load (at-source measures). A second method is to reduce the number of people affected. This can be achieved, for example, by separating functions from each other. Although this does not actually reduce the environmental load, it reduces the height of the bubble and, on balance, enhances the quality of the entire area.

In the introduction, the Amsterdam 'Bubble Method' was described as a method for weighing up alternatives. It is possible to calculate how much each alternative measure contributes in terms of lowering the bubble. The 'yield' of a measure is calculated by measuring the size of the decrease against the budget required for the measure.

Information on the yield of a measure can be used in two ways. First, it can serve as an argument for or against using a certain measure. Second, it can be used to determine the extent to which compensating measures are required.

Apart from a bubble over the whole city, 'secondary bubbles' can be established for different parts of the area (Figure 10.6). If a spatial or environment measure in a given area yields hardly any benefit, other methods are required to enhance the overall quality of the area as a whole.

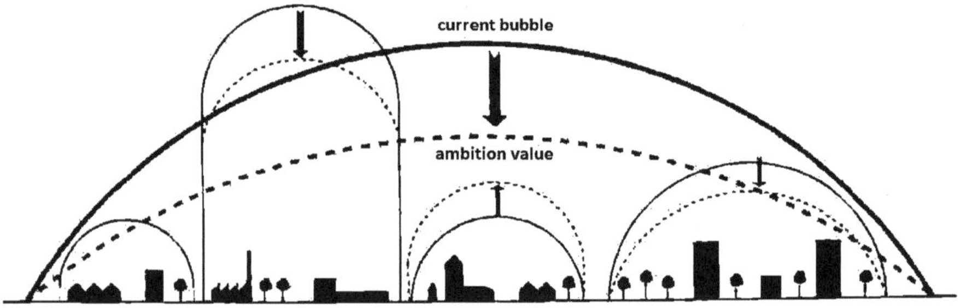

Figure 10.6 Schematic representation of 'secondary bubbles' (Gemeente Amsterdam 1994a)

Practice

It has not proved easy to translate the theoretical principles of the Bubble Method into practical applications. The main aim of the method is to reduce the total scale of the affected area as efficiently as possible. In this process, the nature and size of the load for each location (which are central to standards-based policy) are more or less secondary aspects; the focus is not directly on the relationship between the quality of the physical environment and the source of the load. Hence, it does not focus on the question of whether the standards are actually met. Consequently, the method is relatively far removed from day-to-day policy practice.

In 1996, the Institute for Environmental Studies (IVM) at the VU University Amsterdam reviewed the Bubble Method. This resulted in a shift in the method's theoretical logic, based on what is called 'welfare theory'. This is a theory from economics that is geared towards the conditions that must be fulfilled to maximize the collective prosperity of individuals (De Boer et al. 1996). But the method did not benefit from this 'face-lift'. On the contrary, it seemed to distance it even more from day-to-day policy practice.

For these and other reasons, the Bubble Method has never been used in practice. It proved too abstract and did not provide enough concrete 'handles'. At the time of the review in 1996, the method was tentatively used in two cases: a project at the Southern Axis of Amsterdam and the spatial developments around the Houthavens. This exercise showed that 'it is possible, with limited information, to identify a number of relevant effects, whereby policy alternatives can be compared in terms of collective and individual wealth, and more specifically in terms of economics, liveability, environmental quality and sustainability' (Boer et al. 1996; 76). Nevertheless, the question arises as to how far the method is useful in terms of weighing up options. A great deal of information is required, the calculations are complicated and a relatively large number of assumptions are made.

Discussion

The discussion of the Bubble method has three parts. The first section assesses the position of the method in Dutch environmental policy, based on the framework discussed in Chapter 3. The second part of the discussion involves a more in-depth analysis in which five characteristics will be examined. The third section focuses on the method's scope of application in planning processes.

In comparison to other methods for environmental externalities, the Bubble Method occupies a special position in the framework. The Bubble Method is described as a method that aims to reduce the total environmental load in a given area. This means that it is based not on achieving a single goal but on an overall result. The method can therefore be positioned in the lower section of the framework. The method was designed to help government authorities to weigh up alternatives. There is little or no scope for stakeholder participation. On that basis, the method can be placed in the lower left-hand section of the framework (Figure 10.7a).

This unusual position also becomes evident when we consider the five characteristics. The Bubble Method is a tool for weighing up spatial and environmental options with a view to optimizing the decision-making process. In terms of the characteristic *intention*, the method can be positioned in the middle ring of the model.

The second characteristic is *integration*. We have mentioned the limited extent to which the method has been developed thus far. The framework for weighing up alternatives has been refined for only three environmental aspects. And practice has shown that an 'overall' consideration involving these three aspects is not a simple matter, and leads to calculations such as those for integrated environmental zoning, which were seen as problematic. However, in theory the method has a broader focus, i.e. the benefit yielded by spatial and environmental measures for each bubble. In order to obtain this yield, efforts are focused on considering possible spatial and environmental measures in relation to each other, linking them as effectively as possible and minimizing the cost of operationalizing them. Hence there is definitely integration of the two policy fields. So, despite the initial focus on only three aspects, the method can still be placed in the middle ring in terms of the characteristic *integration*.

The method was designed as a framework that local authorities can use to weigh up alternatives. The process does not allow for participation with other actors. In terms of the characteristic *interaction*, the method can be placed in the innermost ring.

The Bubble Method has various 'secondary bubbles'. Specific goals can be defined for each of these, provided that the overall quality of the area is enhanced. This way of working is known as 'area-specific standardization' (Meijburg and De Knegt 1994). In terms of the characteristic *differentiation*, the method can be placed in the outer ring.

The final characteristic is *conformance and performance*. We do not enter this characteristic of the Bubble Method in the framework. The method is intended to serve as a framework for weighing up alternatives to gain insight into the benefit yielded by possible environmental measures. The Bubble Method does not provide concrete solutions to spatial issues. Figure 10.7b depicts the relationships between the characteristics.

The third part of this discussion focuses on the scope of application of the approach in planning processes. In the discussion above we have emphasized several times that the method is intended for local authorities as a means of weighing up alternatives. Insight into the benefit of certain measures is especially useful at an early stage in a planning process – ideally before the decision-making phase. Another stage at which the information can be relevant is during the monitoring of the implemented policy.

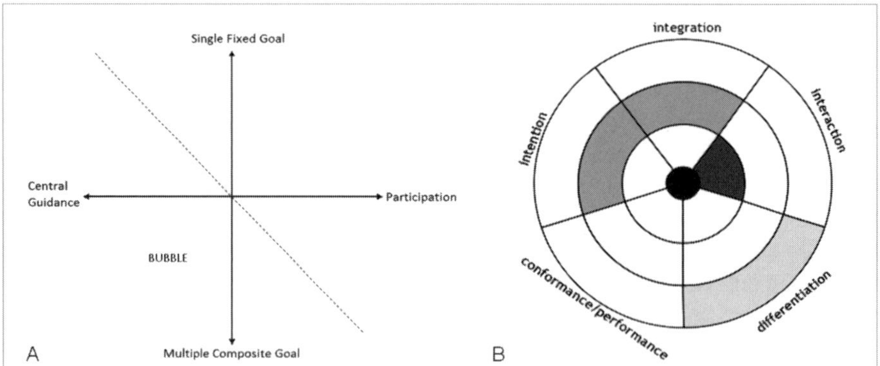

Figure 10.7 Characteristics of the Bubble Method

Conclusion

The survey reveals various strengths and weaknesses. One of the method's strengths is the fact that it focuses on the benefit of intended measures, which can be calculated in advance, albeit with the necessary caveats. In focusing on the consequences of intended measures, the method helps to provide key information at an early stage of the planning process. Second, the method takes account of the sensitive aspects of the various sub-areas. The height of the bubble is determined by factors including the number of people affected by the environmental load.

The method is not yet very strong in terms of how it deals with the principle of compensation. Certain secondary bubbles are easier to reduce than others in terms of their environmental load, and environmental quality can therefore be improved more cheaply and effectively. However, there is a risk that measures will focus on these relatively straightforward areas, and that the difference between the quality of areas will increase. The method does not take account of these growing differences in quality for residents and/or sub-areas, nor does it indicate how much and which form of compensation should be used to address these differences. Another weakness is the calculation method, which is complicated and time-consuming. A great deal of information is required to calculate the index figures. This not only pushes up costs, but can also render the situation less transparent. Finally, it is difficult to translate the principles of the concept into day-to-day practice. The review in 1996 served only to make the method more abstract. It has never come as far as a practical application.

The Bubble Method is an unusual method. Most of the methods for environmental externalities discussed in this book can be positioned on or close to the diagonal of the framework. This diagonal, as we stated in Chapter 3, also shows the development of environmental policy in the Netherlands. The Bubble Method is placed in a completely different part of the framework (Figure 10.8a). Rather than being a method that can be tangibly applied, the main value of the Bubble Method is as a thought model – and not just for the purposes of this book. The method has been of value, for example, in the debate about aligning environmental health and hygiene policy and the economic feasibility of environmental measures for spatial issues. Moreover, the Bubble Method has fed the discussion about the concept of compensation in relation to the traditional system of environmental standards (see De Roo 2001).

10.4 The Other Methods in Perspective

In contrast to the other methods for environmental externalities, the methods in this category cannot be readily compared to each other. They are too dissimilar for this. The Bubble Method remains first and foremost a theoretical thought model, the Environment Maximization Method is a practical design instrument, and the workshop-based approach of the ROMBO tactical tool sets it apart from the other methods. Despite their sometimes exotic approaches, these methods do produce interesting insights – not only for tangible application in spatial issues but also as a source of inspiration for the development of other methods for environmental externalities.

PART C
Relations and Reflections

Chapter 11
The Methods in Relation to Each Other: Going Beyond Descriptions

The range of methods for environmental externalities is multifaceted and diverse. Some of the methods are expressly designed for monitoring input in terms of environmental interests during processes, and others are intended as a medium for providing information or as a 'trade-off' framework for balancing interests. Given the multifaceted nature of the methods, and their sometimes specific applicability, it is necessary to have an understanding not only of each individual method but also of how they relate to each other. For this purpose it is important to compare their background, principles and procedure. This makes it possible to substantiate a decision in favour of a particular method. This information is provided in this chapter.

In Part B of the book, each method is discussed separately. In this chapter, the methods are compared. For this purpose, the comparative model from Chapter 3 can again be used as a guide. The model has three components that enable us to 'zoom in' on the various methods. This process produces a description that varies from 'rough' to 'refined', from 'general' to 'detailed'.

In Section 11.1, the comparison exercise is carried out using the first component of the model. This is used to express the positions of the selected method in Dutch environmental policy. Since this section looks at the whole selection of methods, it is possible to state how the methods relate to each other. The second component of the comparative model is applied in Section 11.2. This component relates to the five defined characteristics: Intention, Integration, Interaction, Differentiation and Conformance & Performance. The final component of the comparative model is central to Section 11.3. It compares the scope of applicability of the methods in spatial planning processes. A conclusion follows in Section 11.4.

11.1 The Methods in Relation to Dutch Environmental Policy

The first component of the comparison is presented as a cross diagram. Each position in the diagram represents a policy choice, a related objective and a constellation of involved actors (De Roo 2004). The diagonal (from top left to bottom right) fulfils an important role in the model. The diagonal represents the optimum circumstances (or conditions) for policy, policy choices and the policy process. The top-left section represents two-down policy; the bottom-right section represents area-oriented approaches. A shift from top-left to bottom-right represents the process of decentralization in Dutch environmental policy. In other words, a shift along the diagonal can be seen as representative of changes in environmental policy.

A shift along the diagonal reflects change in the characteristics and points of departure of policy, and therefore change in how issues are dealt with (Figure 11.1). In Chapter 3 it was argued that the model can therefore help us to understand differences between the categories of methods for environmental externalities and the individual methods. The comparison was based on the discussion of the methods in Part B of the book. Each method was placed in the framework on the basis of characteristics and points of departure.

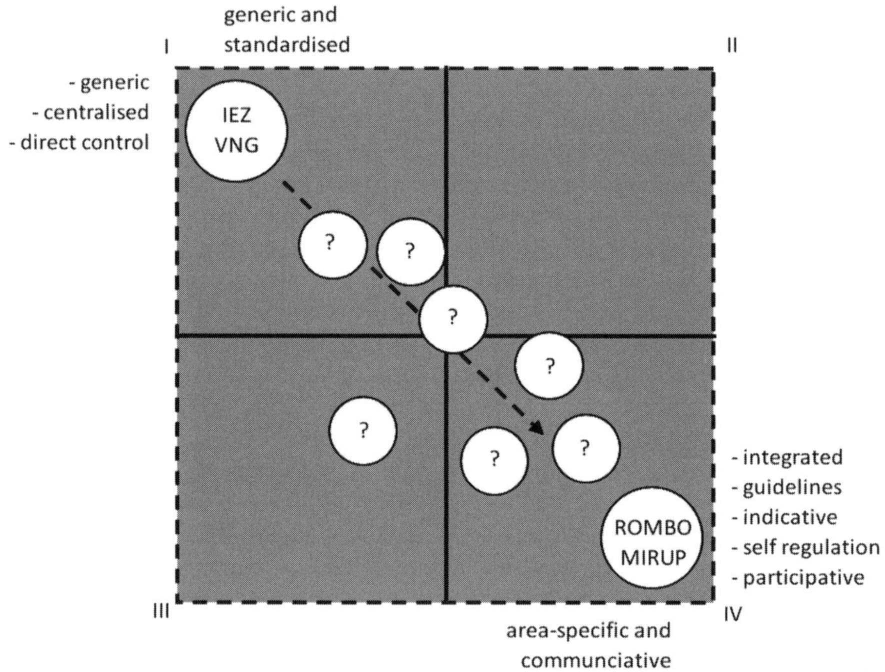

Figure 11.1 Changes in environmental policy, elaborated using characteristics

In total, six categories were distinguished. Some of the categories are close to each other, while others can be seen as each other's opposite extremes (Figure 11.2). The introduction of environmental zoning methods further extended the traditional system of environmental standards. These instruments mainly involve the 'top-down' imposition of framework-setting, generic and above all quantitative standards. The category 'Environmental Zoning Methods' therefore occupies an extreme position, i.e. in the top-left of the model.

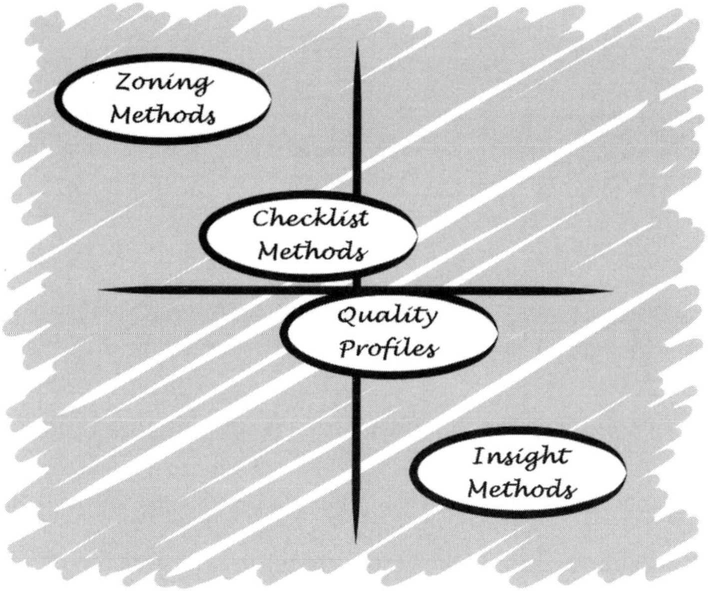

Figure 11.2 Categories in the model for changing environmental policy

The category 'Zip Methods' also occupies an extreme position. These methods offer a perspective for regarding environmental interests as part of the spatial issue. They also emphatically address the matter of identifying stakeholders, who as such must be involved in planning and decision-making. This category can be placed in the lower-right section of the framework

The other categories occupy positions between the two extremes. The Quality Perspectives are closest to the Zip Methods. The main difference between these two categories is the communicative approach advocated by the Zip Methods. There is less emphasis on this in Quality Perspectives. In Checklist Methods too, a broad communicative approach is largely lacking. Furthermore, these methods are characterized by an approach geared to a (technical-rational) policy result that is as far-reaching as possible. The category 'Information Methods' is difficult to position within the model. The position depends to a large extent on how the method is set up and applied (e.g. Environmental Atlases) in policy for the physical environment. The Information Methods could therefore occupy various positions in the framework. There is no single unequivocal position. The category 'Information Methods' is therefore excluded from the diagram.

The categories are positioned in the model on the basis of general characteristics. However, it is quite possible that individual methods occupy a different position to the position of their category as a whole. This is because the positioning of individual methods is based on their specific characteristics. Environmental zoning methods illustrate this. The two methods discussed (VS-IMZ and the VNG method) have been positioned in the part of the framework that corresponds to a generic, framework-setting approach. VS-IMZ (Section 5.2) occupies the most extreme position. This is largely due to the objective-setting nature of environmental policy at the time,[1] the context in which both these methods were developed. Environmental standards set by the national government can stand in the way of weighing-up options at the local level. Later versions of the VNG method (Section 5.1) make greater allowance for differences in the sensitivity of areas. This means that it is possible to weigh up options at local level to a certain extent, resulting in a different position in the framework.

The Checklist Methods were the first methods to allow for a balancing of interests, albeit to a limited extent. Their aim is to incorporate environmental measures in the spatial planning process or to gather information about the quality of environmental health and hygiene of the area concerned. These methods, too, address spatial issues in a quantitative and generic way. Three Checklist Methods are discussed in this book. The Environmental Performance System (Section 6.1) and the *Milieuplaberum* (Section 6.2) – both developed for the Environmental Service in Amsterdam – are methods that allow a balancing of interests to a certain extent. This is less true of Health Impact Screening (Section 6.3) by the GGD (municipal health service), in which the checklist is used mainly to ensure that environmental aspects are not overlooked during the planning process. Moreover, in the Environmental Performance System and Health Impact Screening there is hardly any interaction with the various stakeholders. The *Milieuplaberum* goes a step further in this respect. Its aim is to involve all the relevant local-authority departments in the spatial planning process.

The development of Quality Perspectives created yet more scope for balancing local interests. They also place greater emphasis on interaction with stakeholders. The Bandwidth Method (Section 7.1) of the Utrecht municipal authority was the first method in the 'Quality Perspectives' category. It was developed in the first half of the 1990s, at a time when environmental zoning was still under discussion. In contrast to these zoning methods, the possibilities in a location or area were given priority. Defining the plan quality also introduces self-reflection into the method. This is unique. The characteristics of the Bandwidth Method are perhaps less distinct than those of the other Quality Perspectives. The Neighbourhood Target Scenarios method (Section 7.3) of the Groningen municipal authority thus stands out for its participative approach. Residents and local organizations are involved in the process and can make their wishes known. This communicative approach is not equalled by any of the other Quality Perspectives. The Environmental Quality Profiles method (Section 7.2) of the Maastricht municipal authority, for example, focuses more on the content than the process: area-specific characteristics form the basis for solving spatial issues.

In Chapter 8, we draw parallels between Quality Perspectives and Zip Methods. The Zip Methods provide a 'handle' for the spatial planning process in order

1 Today, too, there is still interest in this 'type' of environmental policy, because it is so clear-cut in terms of what is or is not possible/permitted.

temporarily to incorporate environmental considerations and monitor them during the process. Quality Perspectives and various other methods are part of the 'Zip methods' category. Three Zip methods are discussed in this book. Of these, LOGO (Section 8.2) of the Rijnmond Environment Service is the least focused on process aspects. Rather, its main purpose is to formulate area-specific environmental visions and link these to spatial policy. Process recommendations, as part of LOGO, are included in an appendix. To a large extent, the MILO (Section 8.1) and MIRUP (Section 8.3) approaches of the Hague region (Stadsgewest Haaglanden) are precisely about the planning process. Both methods offer a structure for this, including stages at which environmental interests can be incorporated (e.g. by means of Quality Perspectives). The renewed MILO approach is largely based on the MIRUP tool, which goes a step further in terms of the way in which stakeholders are involved in the process. In the Zip methods there is also a place for methods such as Health Impact Screening and the layer approach.[2]

The methods in the 'Information Methods' category are also difficult to position in the model. Their position depends not only on how the environmental atlases are compiled, but also on how they are applied in policy for the physical environment. This can be done passively, consulting the instrument if a particular question arises, or in an active way. In this case, the instrument is initiated before the planning process begins. One of the purposes of the Groningen Environmental Atlas (Section 10.2) is to involve as many stakeholders as possible when resolving spatial issues. The aim of the Deventer Environmental Atlas (Section 10.1) is primarily to gather relevant information relating to the environment and to use it as input in the planning process. The fact that the two atlases can support the planning process in many ways means that they can be placed in various positions in the framework. As we pointed out above, this is why they are not included in Figure 11.3.

In principle, methods from the category 'Other' can be placed anywhere in the model. The Environmental Maximization Method (Section 9.1) is described as a 'design method'. It is characterized by its interdisciplinary approach and by the fact that no requirements are set in advance with regard to the outcome of the process. The ROMBO tactical tool (Section 9.2) stands out for its almost extreme focus on communication. Finally, we have the Bubble Method (Section 9.3) of the Amsterdam municipal authority. This method also occupies a special position. The Bubble Method was developed as a response and alternative to the generic environmental-zoning methods. Its concept is based on compensation and an area-specific approach to urban and inner-city environmental issues. Hence the Bubble Method is one of the first area-specific (balance) methods.

Figure 11.3 shows the positions of the methods for environmental externalities discussed in this book. The question we can now ask is how can our understanding of the methods be used to substantiate a decision in favour of a particular method?

2 In the discussion of Zip Methods in Chapter 8, we saw how the step-by-step plans regularly refer to external methods. These can be used to support the relevant Zip Method.

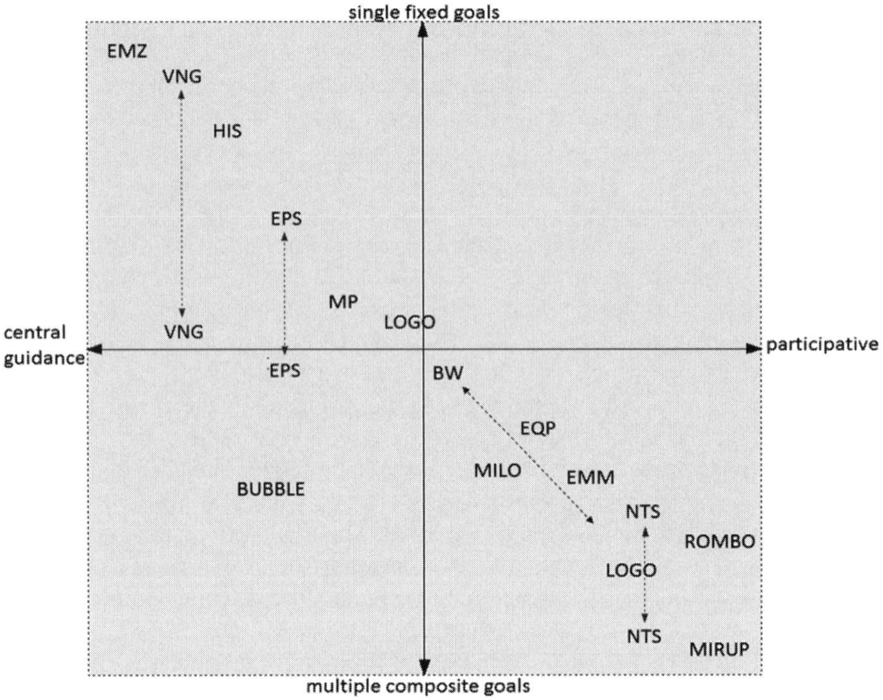

Figure 11.3 Methods for environmental externalities in the model of changing environmental policy

To answer this question, we need to look to the assumptions on which this model is based. In Chapter 3, the model is introduced as a decision model for planning-related choices at various levels of decision-making. The model, as De Roo (2004) argues, can be used to support policy choices by basing arguments on the framework positions.

In a similar way, given the characteristics of the spatial issue, arguments can be formulated to substantiate the choice of one of the methods for environmental externalities. Clearly, then, in the case of issues involving a large number of stakeholders, the appropriate method is one that has a strong focus on participation. Methods of this type are mainly positioned in the lower-right section of the model. This position is associated with many specific features and characteristics. Other characteristics can be considered in a similar way. In the next section, we compare the selected methods on the basis of the characteristics defined in Chapter 3.

11.2 Comparison based on Characteristics

The second part of the comparative model is used to compare the methods for environmental externalities on the basis of five shared characteristics. The model was inspired by the IBO diagram (Section 3.2). The five characteristics are: Intention, Integration, Interaction, Differentiation and Conformance & Performance.

Intention

This characteristic refers to the outcome of the method. Certain methods are based on a system of standards that has a generic result, while others take area-specific characteristics as their starting point, possibly obtaining a completely different result. The two environmental zoning methods discussed (VS-IMZ and the VNG method) are examples of methods that are characterized by generic frameworks and, consequently, a generic outcome. The two methods use a system of standards formulated by the national government. For these methods, the inner ring of the model is shaded, whereby the outcome of the method is assumed to be generic. The Environmental Performance System of the Amsterdam municipal authority can be regarded as a method that, on the one hand, obliges all parties that wish to undertake development projects in a plan area to fulfil a set of minimum requirements. On the other hand, the parties can do this on the basis of their own proposals, given their individual competences and the possibilities available to them in the plan area. In the case of Health Impact Screening, the contours are calculated in accordance with current environmental standards. The Bubble Method works in a similar way: the calculations are based on environmental standards, but the area-specific conditions ultimately determine where policy interventions will or will not be made with regard to environmental quality. After all, it is the 'yield' of the efforts that is important in this method. In cases where the 'yield' is low in terms of an intervention to benefit the environment, compensation or 'substitution' can play a role. The central ring of the model is a good position for these methods. Quality Perspectives, Zip Methods and Information Methods are strongly area-specific, participative or both. In all cases, these methods are adapted to local conditions. Therefore, they can also be positioned in the outer ring of the model. These methods are based on an area-specific approach. Figure 11.4 shows the methods in relation to each other.

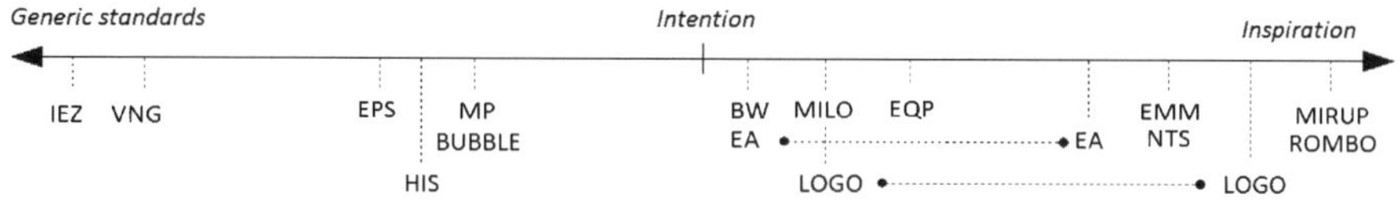

Figure 11.4 Intention

Integration

The integration and coordination of spatial planning and the environment has been a subject for discussion for many years in the policy field for the physical environment. This is an important distinguishing characteristic for methods for environmental externalities. Some methods have a broad orientation that includes virtually every conceivable aspect. By contrast, other methods focus on only a few aspects. The balancing element of the Bubble Method, for example, involves only three environmental aspects. In theory, however, it is possible to include more (environmental) aspects in the Bubble Method. The two environmental zoning methods (VNG and VS-IMZ) are an example of methods that involve the internal integration of environmental aspects. The methods impose environmental requirements on the spatial planning process. Clearly, then, they do not involve external integration (i.e. between the environment and spatial planning). Anyway, all methods actually have an integrated structure to a greater or lesser extent. The Zip Methods, Information Methods and Quality Perspectives are examples of methods that can be described as highly integrated, as can the ROMBO tactical tool and the *Milieuplaberum*. Central positions are occupied by Health Impact Screening, the Environmental Performance System and the Environmental Maximization Method. Figure 11.5 shows the methods in relation to each other.

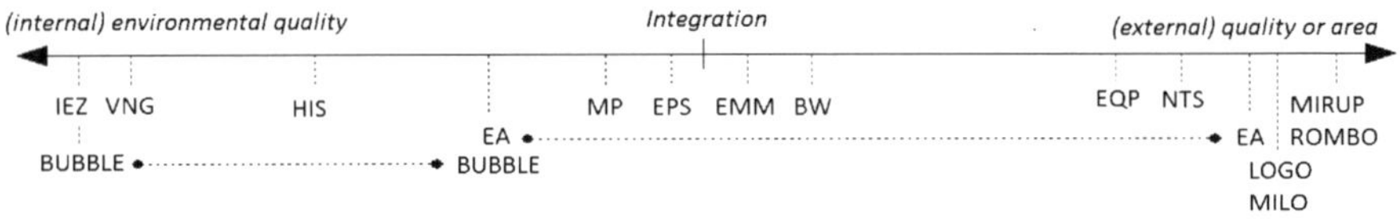

Figure 11.5 Integration

Interaction

In recent decades, the way in which the various actors are involved in planning processes has changed drastically. In the early period of Dutch environmental policy, it was primarily the central government that set the frameworks within which the lower levels of government could implement policy. Today, the government is only one of the several actors involved in policy for the physical environment. Methods for environmental externalities have undergone a similar development. Environmental zoning methods lack almost every form of interaction with stakeholders. This is also true of the Bubble Method and Health Impact Screening. Other methods have greater scope for interaction. These include the Environmental Performance System, the *Milieuplaberum* and the Bandwidth method. Yet the possibilities they offer for interaction are limited (e.g. only a few statutory opportunities to comment). The other selected methods for environmental externalities do make considerable provision for interaction with stakeholders.

Some of these methods were specifically developed with a view to interaction, communication and involvement/engagement. These are the Neighbourhood Target Scenarios, the ROMBO tactical tool, the Groningen Environmental Atlas and MIRUP. Figure 11.6 shows the methods in relation to each other.

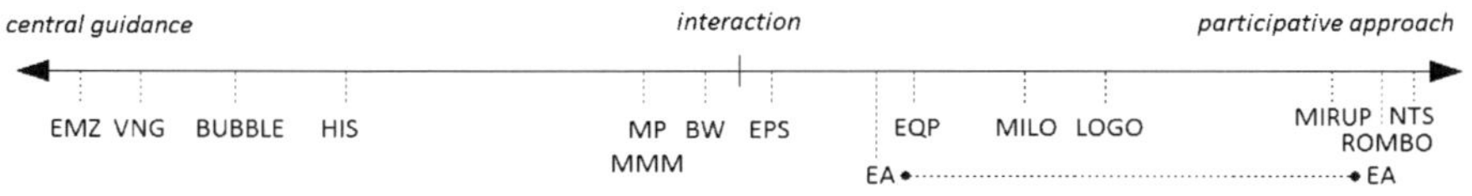

Figure 11.6 Interaction

Differentiation

In recent years, emphasis on the area-specific characteristics of issues has grown. This is reflected in the many area-specific approaches that have been developed. This type of approach is evident in a number of methods for environmental externalities, in particular that of Quality Perspectives. These methods do not involve identifying problems but developing environmental health and hygiene options for the relevant spatial issue on the basis of local and specific characteristics. Examples are the Environmental Maximization Method and the Bandwidth Method. The latter presents the qualities and possibilities of the area as a whole, but offers no scope for further differentiation within the area. These approaches are characteristic of decentralized environmental policy. It is, then, only the most recently developed methods for environmental externalities that enable an area-oriented approach to the spatial issue. Here, we are referring to Quality Perspectives and Zip Methods. The Bubble Method can also be described as an area-specific framework for weighing up alternatives. For the other methods, there are indirectly or to a lesser extent possibilities for differentiation by area type. Figure 11.7 shows the methods in relation to each other.

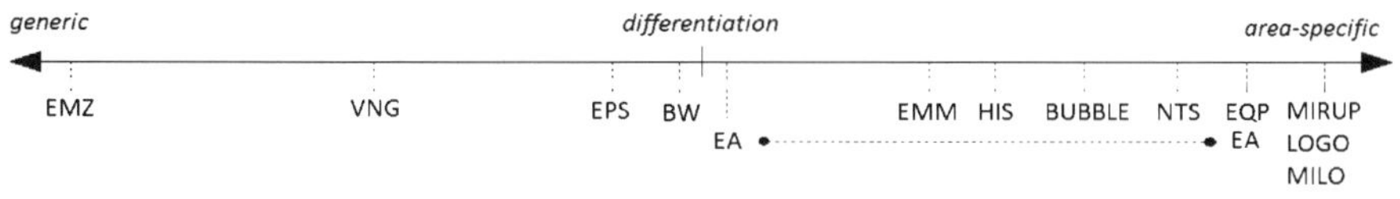

Figure 11.7 Differentiation

Conformance & Performance

Methods for environmental externalities can be applied in various ways. This depends partly on the structure of the method and partly on the party implementing the method. The characteristic 'conformance & performance' relates to how the results of a method influence the eventual policy for the physical environment.

The outcome of the two environmental zoning methods can be described as direct: the environmental zone is directly implemented in spatial policy. The same applies to the Checklist Methods, with the exception of Health Impact Screening. HIS is carried out on a voluntary basis and offers advice 'only'.[3] Other methods have a less direct effect. The least direct methods are Quality Perspectives, Information Methods and Zip Methods, which usually result in a vision for an area, or in ambitions for development, without defining 'hard' objectives. Figure 11.8 shows the methods in relation to each other.

Figure 11.8 Conformance & performance

In this section we compare the selected methods for environmental externalities using the second part of the comparative model: comparison on the basis of inherent characteristics. The picture that emerges is not entirely surprising. In the previous section, the methods were positioned in a model that claims to show the conditions of and changes in Dutch environmental policy. A shift along the diagonal is regarded as a process of decentralization or a development from a centralized, framework-setting approach towards an area-oriented approach. A similar picture emerges from the discussion of the five characteristics. The very first methods (zoning methods) are shown in the top-left section of the model, while more recent participative and area-oriented methods can be found in the lower-right section.

The methods positioned in the top-left section of the model (Figure 11.3) are also, in terms of the five characteristics, associated with a centralized and generic approach. Obviously, the methods in the lower-right section of the model have characteristics in the 'outer ring', associated with participative and area-oriented approaches. A municipal authority looking for a method that enables it to involve local residents in the issue knows which methods may be relevant and which ones it can disregard. In the next section we discuss the scope of applicability of methods for environmental externalities in planning processes.

11.3 Scope for use in Planning Processes

Planning processes for the physical environment are mostly continuous, iterative processes in which it is necessary constantly to weigh up alternatives and make choices. Certain methods were explicitly developed as instruments to support

3 Here too, exceptions prove the rule. In Step 3 of the City & Environment approach, HIS is compulsory and the results must be respected.

decision-making, while others were designed to gather and structure information. This means that the added value of the individual methods is to be found in different stages of the planning process. These differences are explained in the third part of the comparative model for methods for environmental externalities.

The spatial planning process can take various forms. Processes with linear, cyclical and network structures are the most common. A number of different stages are also defined. In Chapter 3, we have chosen to distinguish five stages, for the sake of the overview. The phases are: Initiation, Definition, Design, Decision-making and Monitoring. In this section we compare the methods for environmental externalities in terms of their applicability in the five process stages.

Initiation Phase

The first phase of a planning process focuses on various matters. Information is gathered, for example, on the issue, the characteristics of the plan area and any stakeholders that may be involved. Health Impact Screening is a method that can be used at this stage to obtain information about the relevant aspects of environmental health and hygiene. The two Environmental Atlases can also be used for this purpose. A method that is equally applicable is VS-IMZ, which is an excellent way of gaining insight into the environmental health-and-hygiene situation in the area under consideration. MIRUP and MILO also provide 'handles' for studying the spatial issue and the plan area. The ROMBO tactical tool is appropriate for this phase due to its scope for identifying stakeholders.

Definition Phase

In the Definition phase, the issue is delineated further, thus setting policy goals (plan statements). This may involve, for example, stipulating the types of home to be built or the facilities for the area in question. The desired environmental qualities are also formulated in this phase, in the form of environmental ambitions or otherwise. The Definition phase is an important phase for Quality Perspectives and Zip Methods, since both assume that the ambitions will be defined by the user. The Environmental Performance System is also strongly linked to the Definition phase. This is the phase in which the environmental performance standard is determined so that, in subsequent phases, solutions can be sought that meet the standard. The Definition phase is therefore also the beginning of the solution path for the issue. In theory, for this reason, all the selected methods for environmental externalities can be used to a greater or lesser extent in this phase of the planning process.

Design Phase

In this phase of the planning process, methods for environmental externalities can be used to flesh out the starting points (formulated in the Definition phase) into concrete measures or alternative solutions to the issue. Quality Perspectives and

Zip methods provide 'handles' for this purpose. The Environmental Maximization Method is a tangible design method and can also be used in this phase. The two Environmental Atlases are also appropriate for the Design phase, above all as general support for the design process. The ROMBO tactical tool enables stakeholders to work together to find solutions to the issue.

Decision-making Phase

In the Decision-making phase, various solutions to the spatial issue are weighed up, then a decision is taken. The Bubble Method was one of the first methods that was intended, among other things, to facilitate the weighing up of area-specific aspects with regard to physical interventions. The idea of 'balancing' is also to be found in methods that were developed later. In Chapter 7, the Bandwidth Method is also described as a trade-off method for balancing considerations, and is therefore also appropriate for the decision-making phase. The other Quality Perspectives can also be used to balance the various interests, as can the Zip Methods. The Environmental Atlases can fulfil a general supporting role.

Implementation and Monitoring Phase

This is the final phase of the spatial planning process. In this phase, the proposed solution for the spatial issue is put into practice. This is an important phase for the environment sector. Checks must be made to ensure that environmental aspects have actually been taken into account in the solution to the issue. Largely the same methods are appropriate for this phase as for the Initiation phase: in both phases, information is gathered about the plan area. It should be pointed out that, in this context, the Environmental Atlases do not play an intrinsic role. By contrast, the environmental zoning methods have a range of measurement techniques precisely for this purpose, and are therefore ideal for this phase.

It will always be possible to apply a method in a way other than intended by its originators. In Chapter 5, for example, we discussed the various possibilities offered by the Provisional System for Environmental Zoning (VS-IMZ). The method was embraced in New York, not to enforce the inclusion of environmental considerations in spatial projects but to inform stakeholders about the spatial aspects of the (integrated) environmental quality. The development plans of the Environment Maximization Method, for example, can also be used in the implementation and monitoring phase in order subsequently to assess to what extent the plan statements have been realized. However, this will not be an obvious use in all cases.

11.4 In Conclusion

In this chapter we compared the selected methods for environmental externalities using the comparative model formulated in Chapter 3. The comparison – combined with the discussion of the methods in Part B of the book – should serve as a guide in searching for suitable solutions to certain spatial issues.

An evident conclusion is that there is almost always a suitable method available, especially since the methods discussed in this book cover the whole policy spectrum, from technical-rational (Environmental Zoning methods) to the communicative-rational (e.g. Zip Methods). There is always a method to hand that is suited to the characteristics of a spatial issue. In the communicative part of the spectrum in particular, a wide diversity of methods has been developed. Whichever position a method occupies in the framework, it must be realized that the methods are tools designed to support planning processes geared to spatial developments in a particular way, and in a way that is unique for each method.

Chapter 12
Reflection: What Does it all Mean?

A large number of methods for environmental externalities have been discussed and assessed in this book. Taken together, these methods form a balanced set of instruments for a toolbox that planners can use to maintain or develop a high standard of liveability. But this book is more than that. The methods for environmental externalities are discussed against a background of different and changing circumstances, which are described here in terms of varying degrees of complexity, ranging from straightforward to more complex situations. All these methods together offer a toolbox with a method for basically every situation, while acknowledging that every situation is unique and adjusting measures almost always have to be made to reach the 'perfect' fit.

Environmental policy in the Netherlands and throughout Europe has evolved over the years as the result of an increasing acknowledgement of complexity. The methods for environmental externalities are closely interwoven with this policy context, and, logically, have evolved in step with it. This co-evolution has led to a set of methods for environmental externalities whereby a useful approach is available for almost every situation.

The structure of Dutch environmental policy in the 1970s was the basis for the development of generic, framework-setting conditions in the 1980s. These conditions became the quantitative and generically applicable environmental standards that set limits on spatial development relating to environmentally intrusive functions. At the beginning of the 1990s, the first doubts arose regarding this path. In the mid-1990s, there was talk of the 'fall of the standard'. 'Changing environmental policy' is not a recent phenomenon. Since the beginnings of structural environmental policy in the Netherlands, the policy field has always been in flux.

From the discussion of the selected methods for environmental externalities, we can deduce that these methods and techniques have, to a certain extent, kept pace with the developments in Dutch environmental policy. This leads not only to the observation that each period in environmental policy has produced appropriate methods for environmental externalities, but also to the observation that a specific method exists for almost every type of issue.

12.1 Methods for Environmental Externalities: A Question of Customization

In a period of about forty years, Dutch environmental policy has gone through various stages of development, including major achievements such as the first

National Environmental Policy Plan (*Nationaal Milieubeleidsplan*) and various low points, such as the largely unsuccessful attempts to integrate environmental policy and spatial policy. The instruments, in the form of methods and techniques, have evolved in step with this development. They proved to be methods and techniques that can help to provide greater insight into the environmental health and hygiene aspects of spatial issues, and offer solutions or an approach to maintain – and where possible improve – liveability. The methods for environmental externalities can serve as a framework for planning processes and for spatial development, but equally as a guidance instrument for achieving a high standard of liveability.

In the 1980s and early 1990s, the methods were largely geared to assessing and testing the current environmental situation. The aim of these early methods was primarily to offer appropriate solutions to the environmental/spatial problem. This was done mainly with rigid frameworks for physically separating environmentally sensitive and environmentally intrusive land uses. In subsequent years the methods became more quantitative in nature, partly due to the influence of changing insights in environmental policy.[1] It is no longer a matter of assigning a quantitative score to a particular situation but, for example, of formulating proposals for the external integration of environmental policy and policy on spatial development, or stimulating communicative interaction between actors in the field of environmental policy and actors in the field of spatial development.

The first methods for environmental externalities, known initially as 'environmental assessment methods', can be described as 'testing' methods The criteria relating to the content of these methods are usually determined in advance by the government, either with or without statutory norms. The Provisional System for Integrated Environmental Zoning (VS-IMZ, *Voorlopige Systematiek voor Integrale Milieuzonering*) can be regarded as the ultimate example of this. This proved to be a method whereby the government makes use of direct governance.

The later methods for environmental externalities are very different. Although there will always be exceptions, the methods developed recently generally have a communicative approach and intelligent quality differentiations are used to find 'customized' solutions. To an important extent, these methods are based on participation with other actors and the local support base. These methods can be seen much more as *guidance instruments* for spatial planning. Examples are LOGO and MIRUP, as well as Neighbourhood Target Scenarios and the ROMBO tactical tool. It is no longer the government that defines the content-related criteria and policy frameworks through direct governance. The actors involved work together to determine the solution strategy in these methods. The government participates as one of the actors, or takes on a facilitating role. There is indirect governance or self-governance rather than direct governance. As a result of the developments in environmental policy described above, responsibility for environmental policy has increasingly been shifted to local and regional authorities. These developments do

1 As always, there are exceptions. Health Impact Screening, for example, is still a method for quantifying a given situation from the perspective of health (see also Section 6.3).

not lead to a 'complete' rounded conclusion (see De Roo 2004 for a discussion), which leaves open the question of what is to come.

We can also conclude that the methods developed recently have more in common with each other than the methods that were developed in the 1990s. This is partly due to the fact that they are largely based on each other. Of the selection discussed here, between 5 and 8 methods can be regarded as Quality Perspectives. But there is also substantial divergence between these recently developed methods. Some methods are based on a more communication-oriented approach, while others are geared to solutions that strike a balance between various local considerations. The selection of methods for environmental externalities discussed here shows such a pattern (Figure12.1).

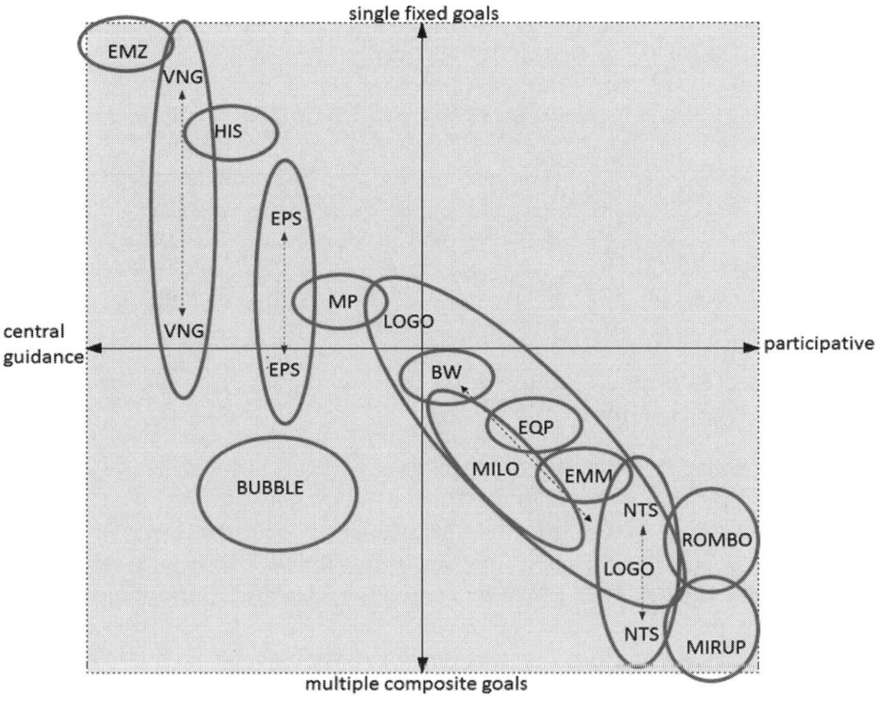

Figure 12.1 The range in the development of methods for environmental externalities

At first sight, the methods appear to occupy a random position in the framework. However, the position that a particular method occupies in the framework tells us about its characteristics (see also Chapter 3). Notably, the methods are, to a certain extent, concentrated around an imaginary diagonal that extends from top left to bottom right. Within a certain bandwidth of this diagonal, policy and decision-making are assumed to be optimal (De Roo 2004, see also Chapter 3 and Figure 12.2).

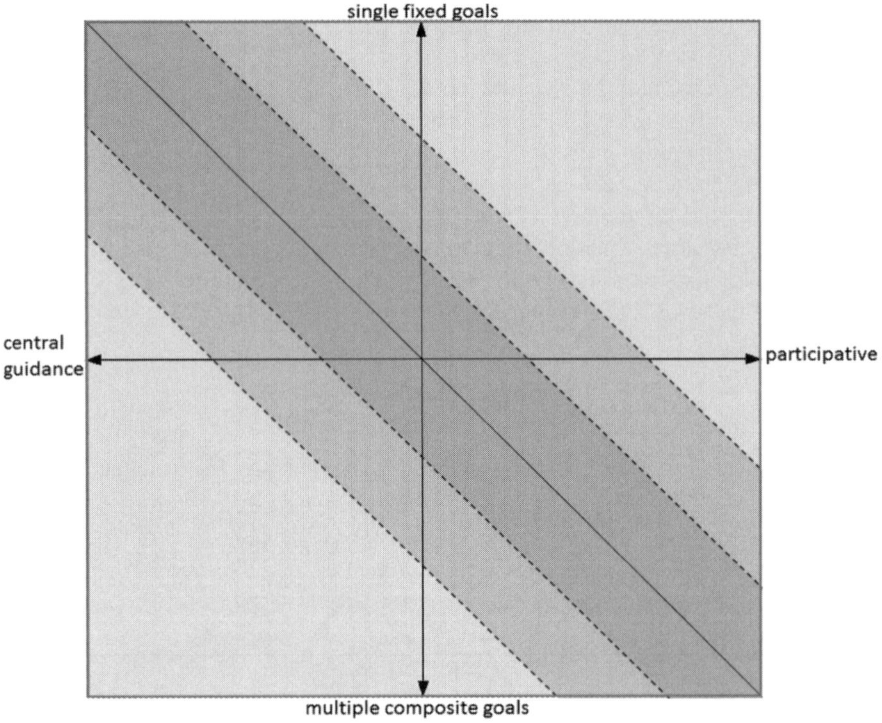

Figure 12.2 Optimum bandwidth for methods for environmental externalities

Environmental zoning methods can be regarded in the same way. In this respect, VS-IMZ and the VNG method occupy optimum positions. We can expect considerable coherence between the characteristics of an issue and the specific features of the method for environmental externalities – at least, when dealing with the 'right' type of issue, which can be positioned in the framework in a similar way to the method.[2]

The distribution of the many methods for environmental externalities within the optimum bandwidth makes it possible to link issues in the framework to the most appropriate method. A method that can be regarded as sub-optimal or even minimal is the 'Bubble Method' (*Stolpmethode*) of Amsterdam's municipal authority.Not surprisingly, this method has never been put into practice.

2 In Chapter 3, we link the framework to the level of complexity as a decision-making criterion. The issues positioned in the top left of the framework are straightforward; those positioned in the lower right of the framework are highly complex. Complex issues occupy positions somewhere in the middle. From this we can conclude that VS-IMZ and the VNG method can only be applied to straightforward issues.

12.2 Smart Methods do not come Easy

We have observed, but not discussed, the fact that local authorities are largely unaware of the possibilities that methods for environmental externalities offer in terms of supporting local policy for the physical environment.[3] The MILO project bureau has attempted to raise awareness by organizing workshops on the MILO approach. The original tool was an insufficient 'handle' for local authorities. The method was complicated and not very transparent. It can thus be compared to VS-IMZ, which disappeared from the scene for similar reasons. Unlike VS-IMZ, the potential of a MILO approach is not seen in terms of a 'complete' final outcome (the sustainable physical separation of environmentally intrusive and sensitive land uses) but in terms of its possibilities (a range of options and 'handles'). This results in the development of flexible, hybrid methods, as visible in the MILO approach.

MILO (see Section 4.4) dispenses with the original seven-jump 'dance' from similar methods such as MIRUP and LOGO. Instead, techniques from different methods for environmental externalities are combined and applied in order to assure liveability in spatial projects or when designing policy for the physical environment.

> '...and then you see people shaking hands with each other...'
> 'We give workshops on MILO all over the Netherlands. We visit large and small municipalities, and municipalities that have experience with area-specific approaches and those that don't. It is shocking to see staff from different departments shaking hands and introducing themselves for the first time. Environmental and Spatial Planning departments still operate very much in parallel, and are not – or not sufficiently – aware of each other's work. Until this changes, suspicion and frustrations will persist. We hope that, through the workshops, these people will learn about the possibilities of the MILO approach.'*
>
> * Interview with Van Liefland 05–07–2006.

This book about methods for environmental externalities can play a supporting role in communicating the body of ideas behind MILO. It is therefore not intended as a step-by-step plan for choosing and following a method. The main aim of this book is to introduce the possibilities that methods for environmental externalities offer in terms of supporting policy for the physical environment. The methods chosen in this book have something for everyone: from methods for evaluating to externality methods, and from information methods to methods that support processes.

However, these methods cannot simply be applied straightaway. There is work that has to be done first. Some methods are relatively simple and require little work (e.g. the checklist methods), but others are more complicated and require more

3 Interview with Hanegraaf and Van Helden 05–07–2006.

knowledge and resources (e.g. Quality Perspectives). An important precondition for using such methods is that sufficient up-to-date information is available.[4] In that context, we have already referred to the possibilities for preparatory work that Environmental Atlases offer.

Finally, a remark about implementing methods in policy for the physical environment. The municipal authorities of Maastricht and Zwijndrecht have shown that implementing Environmental Quality Profiles in local environmental policy plans or structural concepts can be very successful. This is also true in the case of Apeldoorn (Environmental Quality Profiles) and Groningen (Neighbourhood Target Scenarios). The methods for environmental externalities applied in the Netherlands have now proved their worth, but are still relatively isolated in relation to spatial-development policy. A link with spatial-development plans, such as a structural concept or zoning plan (as in Amsterdam's Environmental Performance System) can help to integrate spatial development and liveability.

12.3 Finally

In the first place, the methods for environmental externalities developed in the Netherlands show creativity and innovation. A range of methods is available – reflecting shifts in policy away from central governance towards a local 'tailored' approach – that enable the user to find an appropriate method for almost every situation. This perspective, set out in 'Smart Methods' is unknown or little known to users, particularly the local authorities. It is important, in terms of environmental policy and the quality of the physical environment, to inform potential users about this wealth of options. It is then down to the users to adapt the various possibilities to their own situation. In that context we refer to the MILO approach, which involves applying techniques from various methods in order to support policy for the physical environment.

'Smart Methods' for environmental externalities show that Dutch methods dealing with spatial effects of environmental intrusion and environmentally sensitive functions can set limits on, and serve as a tool for, environmental-spatial conflicts. These methods can also serve as an inspiration for developing new methods and techniques. Although Dutch national environmental policy lost its prominent and progressive position in the international field some time ago, the developments in the area of 'smart methods' for environmental externalities certainly belong to the 'pride of Holland'.

4 Interview with Van Liefland 05–07–2006.

Bibliography

Akkersdijk, I.G. (2003) Gezondheidseffect Screening (GES) bij stedenbouwkundig plan Gezondheidspark Dordwijk te Dordrecht, GGD Zuid-Holland Zuid, Dordrecht.

Ale, B. (2003) *Ons overkomt dat niet*, inaugurele rede, Faculteit Techniek, Bestuur en Management, Technische Universiteit Delft, Delft.

Anderson, N., E. Hanhardt, I. Pasher (2005) From Measurement to Measures; Land Use and Environmental Protection in Brooklyn, New York, in: Miller, D., G. de Roo (eds) *Urban Environmental Planning; Policies, Instruments and Methods in an International Perspective*, Ashgate, Aldershot, pp. 35–40.

Arcadis Heidemij Advies (2001) *Wijkstreefbeelden Leefmilieukwaliteit*, Gemeente Apeldoorn, Arcadis, Arnhem.

Bakker et al. (1999) *Duurzaam Bouwen in Den Haag*, Villerica Publishing, DSO gemeente Den Haag.

Bakker et al. (2004) *Projectenboek Duurzaam Bouwen*, DSO gemeente Den Haag.

Bakker, H.J., P.P.J. Driessen, J.J. van den Berg (1998) *Afstemmen en instemmen; Ruimtelijke ordening en milieubeheer, integratie en coördinatie van het omgevingsbeleid op alle overheidsniveaus*, VUGA Uitgeverij, Den Haag.

Bartelds, H., G. de Roo (1995) *Dilemma's van de compacte stad; Uitdagingen voor het beleid*, VUGA Uitgeverij, Den Haag.

Berge, A. van den, M. Brouwer, K. Duijvestein, A. Hoogers, A. Wijnants (1998) Milieu Maximalisatie Methode; Structurerende inbreng van milieu in stedenbouwkundige planvorming, BOOM-Duijvestein, Delft.

Blanco, H. (2004) Lessons from an Adaption of the Dutch Model for Integrated Environmental Zoning (IEZ) in Brooklyn, NYC, in: D. Miller, G. de Roo (eds) *Integrating City Planning and Environmental Improvement; Practicable Strategies for Sustainable Urban Development*, Ashgate, Aldershot, pp. 135–54.

Boei, P.J. (1993) Casus Arnhem; Integrale milieuzonering op en rond het industrieterrein Arnhem-Noord, in: G. de Roo (red.), *Kwaliteit van norm en zone; Planologische consequenties van (integrale) milieuzonering*, Geo Pers, Groningen, pp. 75–82.

Boer, J. de, V.M. Sol, F.H. Oosterhuis, J.F Feenstra, H. Verbruggen (1996) *De Stadsstolpmethode; Een afwegingskader voor de integratie van milieu, economie en ruimtelijke ordening bij stedelijke ontwikkeling*, Instituut voor Milieuvraagstukken, Vrije Universiteit, Amsterdam.

Borst, H. (1996) Integrale Milieuzonering; Ontwikkelingen, consequenties en mogelijkheden, in: G. de Roo (red.) *Milieuplanning in vierstromenland*, Samsom H.D. Tjeenk Willink, Alphen aan den Rijn, pp. 94–108.

Borst, H., G. de Roo, H. Voogd, H.van der Werf (1995) *Milieuzones in beweging; Eisen, wensen, consequenties en alternatieven*, Samsom H.D. Tjeenk Willink, Alphen aan den Rijn.

Bouwer, K. (1998) Het spanningsveld tussen milieu en ruimtegebruik, in: Bakker, H.J., P.P.J. Driessen, J.J. van der Berg (red.) *Afstemmen en instemmen; Ruimtelijke ordening en milieubeheer, integratie en coördinatie van het omgevingsbeleid op alle overheidsniveaus*, VUGA Uitgeverij, Den Haag, pp. 11–22.

Brimblecombe, P., F. Nicholas (1995) Urban air pollution and its consequences, in: T. O'Riordan (red.) *Environmental Science for Environmental Management*, Longman Scientific & Technical, Harlow (UK), pp. 283–95.

BRU-projectgroep (Bestuur Regio Utrecht) (1995) Bandbreedte in beeld, handreiking planvorming en bestuurlijke beoordeling, testcase milieukwaliteitsbeeld Leidsche Rijn, Regionaal Beraad Utrecht, Utrecht.

BRU-projectgroep (Bestuur Regio Utrecht) (1996a) Bandbreedte; Integraal afwegings-kader, Rapportage over de 2e fase, BRU, Utrecht.

BRU-projectgroep (Bestuur Regio Utrecht) (1996b) Bandbreedte; Integraal afwegings-kader; Milieukwaliteitsbeeld bestemmingsplan Zuilen, BRU, Utrecht.

Carson, R. (1962) *Silent Spring*, Houghton Mifflin, Boston.

CLTM (Commissie Lange Termijn Milieubeleid) (1990) *Het Milieu; Denkbeelden voor de 21ste eeuw*, Kerckebosch bv, Zeist.

Commissie De Wolff (1970) Rapport van de Commissie voorbereiding onderzoek toekomstige maatschappij-structuur, Staatsuitgeverij, Den Haag.

Dammers, E., F. Verwest, B. Staffhorst, W. Verschoor (2004) Ontwikkelingsplanologie; Lessen uit en voor de praktijk, Ruimtelijk Planbureau, NAi Uitgevers, Den Haag.

DCMR Milieudienst Rijnmond (2003) Milieubeleidsplan 2003–2007; Onderweg naar een duurzaam Krimpen aan den IJssel, DCMR, Schiedam.

DCMR Milieudienst Rijnmond (2004) *Handreiking Locale Gebiedstypologie en Omgevingskwaliteit; Naar een duurzame ruimtelijke kwaliteit*, DCMR, Schiedam.

DCMR Milieudienst Rijnmond (2005) *Milieuatlas Stormeiland; Advies met betrekking.*

Douma, W.Th., K. Pieters, K. Feenstra, R. Koch-Hartmanová (2007) Pilot-Monitor EU-invloed; Een onderzoek naar de realiseerbaarheid van een Permanente Monitor voor het meten van de invloed van Europese regelgeving op in Nederland geldende wet- en regelgeving, T.M.C. Asser Instituut, Den Haag.

EC (Europese Commissie) (2003) European Common Indicators; Towards a Local Sustainability Profile, Final Project Report, Ambiente Italia Research Institute, Milaan.

Eijk, P.J. van, A.J. Dijkstra, S. Hermans, S.P. Tjallingii (2000) Naar een duurzaam Poptahof in de 21ste eeuw. Een verkenning van de kansen, DIOC-DGO, Delft.

Elzinga, D.J., G.H. Hagelstein (1998). 'Centralisatie en decentralisatie', in: A.F.A. Korsten en P.W. Tops (red.), *Lokaal bestuur in Nederland. Inleiding in de gemeentekunde*, Alphen aan de Rijn, Samson, p. 111.

Endenburg, G. (2001) *Sociocratie als sociaal ontwerp in theorie en praktijk*, Eburon, Delft.
Environment; The Renewal of Environmental Policy Arrangements', Dordrecht.
EU (Europese Unie) (2001) Richtlijn 2001/42/EG van het Europees Parlement en de Raad betreffende de beoordeling van de gevolgen voor het milieu van bepaalde plannen en programma's, Brussel.
Evertse, M., J. van Grootheest, J. Sikkema, H. Vlek (2003) Milieu en Methode; Een onderzoek naar milieubeoordelingsmethoden, leeronderzoek, Faculteit der Ruimtelijke Wetenschappen, Rijksuniversiteit Groningen.
Fast, T. (2002) *Evaluatie van de GES Stad&Milieu en het Steunpunt GES*, Fast Advies, Utrecht.
Fast, T., P.J. van den Hazel, D.H.J. van de Weerdt (2004) Gezondheidseffect Screening Stad&Milieu; Handboek voor een gezonde inrichting van de woonomgeving, GGD Nederland, Utrecht.
Feenstra, S., M. de Wever (1998) Bandbreedte Integraal Afwegingskader; Monitoren van milieukwaliteit, Bestuur Regio Utrecht, Utrecht.
Fleurke, F., R. Hulst, P.J. de Vries (1997) *Decentraliseren met beleid*, Sdu, Den Haag.
Geel, P. van (2004) Speech van staatssecretaris Van Geel van VROM op de Dag Stad en Milieu/MILO, 13 mei 2004, Rotterdam.
Gemeente Amsterdam (1991) Nota van Uitgangspunten voor de IJ-oevers; Amsterdam naar het IJ, gemeente Amsterdam, Amsterdam.
Gemeente Amsterdam (1994a) Beleidsnota Ruimtelijke Ordening en Milieu, dienst Ruimtelijke Ordening en Milieudienst, Amsterdam.
Gemeente Amsterdam (1994b) Bestemmingsplan IJ-oevers, dienst Ruimtelijke Ordening, Amsterdam.
Gemeente Amsterdam (1997) Milieuplaberum; Milieuchecklist voor stedenbouwkundige plannen in Amsterdam, gemeente Amsterdam, Amsterdam.
Gemeente Amsterdam (1999) De Gids voor Ruimtelijke Ordening en Milieu; Gereedschapsgids, dienst Ruimtelijke Ordening en Milieudienst, drukkerij Mart Spruit bv, Amsterdam.
Gemeente Amsterdam (2006) Plan- en Besluitvormingsproces Ruimtelijke Maatregelen (Plaberum), in: *Bouwbrief; Regels en afspraken Amsterdamse woningbouw*, nummer 2006–38, Amsterdam, pp. 1-2.
Gemeente Amsterdam & Google Maps (2009) Ruimtelijke vertaling van klachtenregistraties in Geuzenveld-Slotermeer, http://mor.amsterdam.asp4all.nl/MORGeuzenveld.aspx, (bezocht op 23–04–2009).
Gemeente Deventer (2003a) Milieuatlas; Het milieu op de kaart, Milieubeleidsplan 2003–2008, Sector Ruimte, Milieu en Wonen, Deventer.
Gemeente Deventer (2003b) Milieuvisie 2003–2008; Het milieu op de kaart, Sector Ruimte, Milieu en Wonen, Deventer.
Gemeente Deventer (2005) Milieuatlas; Het milieu op de kaart, geactualiseerde kaarten, Sector Ruimte, Milieu en Wonen, Deventer.

Gemeente Groningen (1993) Handleiding Milieubeoordelingsmethode Groningen, Milieudienst, Groningen.

Gemeente Groningen (2001) Milieubeleidsplan 2001–2004; Lokaal Gewogen, Milieudienst, Groningen.

Gemeente Groningen (2005) Milieubeleidsplan 2005–2008; Stadsleven Natuurlijk, Milieudienst, Groningen.

Gemeente Haarlemmermeer (2003) Milieukansenkaart Haarlemmermeer 2030, gemeente Haarlemmermeer, Hoofddorp.

Gemeente Hellvoetsluis (2002) Milieustructuurplan 2002–2006, Productgroep Milieu gemeente Hellevoetsluis, Hellevoetsluis.

Gemeente Maastricht (2001) Natuur- en milieuplan Maastricht 2030; Niet overal een beetje!, Gemeente Maastricht, Maastricht.

Gemeente Rotterdam (1998) Milieu op z'n Plek; Milieukompas voor plannenmakers en bestuurders, Drukkerij De Eendracht, Schiedam.

Gemeente Vlaardingen (2004) Structuurplan Rivierzone, gemeentelijk projectbureau Rivierzone, Vlaardingen.

Gemeente Zwijndrecht (2003) Milieubeleidsplan Zwijndrecht 2003–2006; Samen werken aan een leefbaar en duurzaam Zwijndrecht, gemeente Zwijndrecht, Zwijndrecht.

GGD Nederland (Gemeentelijke Gezondheidskundige Dienst) (2004) *Gezondheidseffect Screening Stad&Milieu; Voor een gezonde inrichting van de woonomgeving*, Tailormade, Buren.

Hardin, G. (1968) *The Tragedy of the Commons*, in: Science, vol. 162, pp. 1243–48.

Hidding, M. (1997) *Planning voor stad en land*, Uitgeverij Coutinho, Bussum.

Hidding, M.C., M. Kerstens (2001), Omgevingsplanning, een verkenning. Over de aard en reikwijdte van een nieuw planologisch begrip, in: G. de Roo, M. Schwartz (red.), *Omgevingsplanning, een innovatief proces*, Sdu, Den Haag.

Hoogerwerf, A., M. Herweijer (1998), *Overheidsbeleid: een inleiding in de beleidswetenschap*, Alphen aan de Rijn, Samson.

Humblet, A.G.M., G. de Roo (red.) (1995) *Afstemming door inzicht; Een analyse van gebiedsgerichte milieubeoordelingsmethoden ten behoeve van planologische keuzes*, Geo Pers, Groningen.

IPO (Interprovinciaal Overleg) (2007) PRISMA 2007; Programma Ipo Strategische Milieu Agenda, IPO, Den Haag.

Kamphorst, D.A. (2006) *Veranderend Milieubeleid; Een onderzoek naar decentralisatie, doorwerking en integratie van milieubeleid in een stedelijke context*, Geo Pers, Groningen.

Kasteren, J. van (1985) 15 jaar milieubeleid; Geluid, in: *Intermediair*, nr. 48, pp. 33–5.

Kickert, W.J.M. (1986), *Overheidsplanning, theorieën, technieken en beperkingen*, Van Gorcum, Assen/Maastricht.

Klaassen, A.W. (1994) *Ruimtelijk beleid in theorie en praktijk*, VUGA Uitgeverij, Kluwer Academic Publishers ,Den Haag...

Lange, M. de (1995) *Besluitvorming rond strategisch ruimtelijk beleid; Verkenning en toepassing van doorwerking als beleidswetenschappelijk begrip*, Thesis Publishers, Amsterdam.

Linden, J.F. van der, J. Visser, C. Zuidema (2001) *Kwartetten om milieuhygiëne; Een aanzet tot een milieuhygiënische atlas*, leeronderzoek, Faculteit der Ruimtelijke Wetenschappen, Rijksuniversiteit Groningen, Groningen.

Loo, B. van der, M. van Bruggen (1999) Gezondheidseffectscreening milieu en gezondheid; Fase 2: test GES Stad en Milieu, Landelijke Vereniging van GGD en RIVM, Utrecht.

Lubach, D.A. (1991) De juridische betekenis van de beschrijving in hoofdlijnen, in: P.J.J. van Buuren (red.) *25 jaar WRO*, Kluwer, Deventer, pp. 73–81.

Meadows, D., J. Randers, W.W. Behrens III (1972) *The Limits to Growth*, Universe Books, New York.

Meijburg, E., M. de Knegt (1994) Een stolp over Amsterdam; milieu en ruimtelijke ordening verstrengeld, *ROM*, nr. 10, pp. 9–12.

Michiels, F.C.M.A. (1989) Wet inzake de luchtverontreiniging, in: Brussaard, W., G.H. Addink (red.) *Milieurecht*, W.E.J. Tjeenk Willink, Zwolle.

Miller, D. (2004) Design and Use of Urban Sustainability Indicators in Physical Planning; A view from Cascadia, in: D. Miller, G. de Roo (eds) *Integrating City Planning and Environmental Improvement; Practicable Strategies for Sustainable Urban Development*, Ashgate, Aldershot, pp. 245–66.

MNP (Milieu- en Natuurplanbureau) (2004) *Milieubalans 2004; Het Nederlandse milieu verklaard*, MNP/RIVM, Bilthoven.

Oosterhoff, H., G. de Roo, M.J.C. Schwartz, H. van der Wal (2001) Omgevingsplanning in Nederland; Een stand van zaken rond sectoroverschrijdend, geïntegreerd en gebiedsgericht beleid voor de fysieke leefomgeving, onderzoek in opdracht van de Rijksplanologische Dienst, VROM, Den Haag.

Oosterlee, A., R.H. Keuken (2004) Gezondheidseffecten van luchtverontreiniging door fijn stof in de IJmond, de bijlagen, GGD Kennemerland, Haarlem.

Oosting, M (red.)(1984), Aspecten van decentralisatie, Staatsuitgeverij, Den Haag.

OpdenKamp Adviesgroep (2004) Project Implementatie Aarhus (PRIMA) – Milieu-informatie, rapport FASE 1, OpdenKamp Adviesgroep bv, Den Haag.

Ottens, E. (2001) Een brug tussen milieu- en ruimtelijk beleid, in: *ROM*, nr. 11, pp. 34–5.

Poll, R. van (1997) *The perceived quality of the urban residential environment; A multi-attribute evaluation*, Westrom Drukkerij, Roermond.

Provincie Zeeland (2006) *Zeeuwse leefomgeving, prov. zeeland, www.zeeland.nl/loket/milieuregister* (bezocht in december 2006).

Raad van State (1997) Uitspraak in de zaak bestemmingsplan 'Noordschil Bedrijventerrein', zaaknummer E01.94.0433, 13–05–1997, Afdeling Bestuursrechtspraak van de Raad van State, Den Haag.

Raad van State (2002), Voorstel van wet met memorie van toelichting houdende nieuwe regels omtrent de ruimtelijke ordening (Wet ruimtelijke ordening),

Kamerstukken II 2002/03, 28 916, nr A., 12–07–2002, Raad van State, Den Haag.

Raad van State (2006) Uitspaak in het hoger beroep van 'Victoria Winterswijk BV', zaaknummer 200605602/1, 06–12–2006, Afdeling Bestuursrechtspraak van de Raad van State, Den Haag.

Ravesloot, C.M. (2002) Controlled Creativity with Maximum Feasibility; Organising Sustainable Building in the City of The Hague, in: Buijs, J., R. van der Lugt, H. van der Meer (eds) *Idea Safari; Creativity and Innovation*, Twente University Press, Enschede.

Ravesloot, C.M. (2005) *Rombo-tactiek; Ontwikkeling van een organisatiemethode voor realisatie van energieneutrale woningbouw in Nederland*, Bouwstenen Publikatiebureau, Eindhoven.

RIVM (Rijksinstituut voor Volksgezondheid en Milieuhygiëne) (1988) *Zorgen voor morgen; Nationale milieuverkenning 1985–2010*, Samsom H.D. Tjeenk Willink, Alphen aan den Rijn.

RMO (Raad voor Maatschappelijke Ontwikkeling) (2003) Bevrijdende *kaders; Sturen op verantwoordelijkheid*, advies 24, RMO, Den Haag.

Roo, G. de (1999, 2001) *Planning per se, planning per saldo; Over conflicten, complexiteit en besluitvorming in de milieuplanning*, Sdu Uitgevers, Den Haag.

Roo, G. de (2002) *De Nederlandse Planologie in Weelde gevangen; Van ruimtelijk paradijs naar een leefomgeving in voortdurende staat van verandering...*, oratie, Faculteit der Ruimtelijke Wetenschappen, Rijksuniversiteit Groningen, Groningen.

Roo, G. de (2004) *Toekomst van het milieubeleid; Over de regels en het spel van decentralisatie: een bestuurskundige beschouwing*, Koninklijke Van Gorcum, Assen.

Roo, G. de, H. Voogd (2004) *Methodologie van planning, over processen ter beïnvloeding van de fysieke leefomgeving*, Coutinho, Bussum.

Roo, G. de, M. Schwartz (red.) (2001) *Omgevingsplanning, een innovatief proces; Over integratie, participatie, omgevingsplannen en de gebiedsgerichte aanpak*, reeks Planologie deel 2, Sdu Uitgevers, Den Haag.

Rosdorff, S., L.K. Slager, V.M. Sol, K.F. van der Woerd (1993) *De Stadsstolp; Meer ruimte voor milieu en economie*, Instituut voor Milieuvraagstukken, Vrije Universiteit, Amsterdam.

Sabatier, P. A., H. C. Jenkins-Smith (eds) (1993) *Policy Change and Learning; An Advocacy Coalition Approach*, Westview Press, Boulder, CO.

Sawicki, D., D. Peterman (2002) Surveying the extent of PPGIS practice in the United States, Hoofdstuk 2 in W. Craig, T. Harris and D. Weiner (eds), 2002, *Community Participation and Geographic Information Systems,* London: Taylor and Francis, p. 17–37.

Schmit, H. (2005) *Frambozenschol en Bloemkoolpaling; Nederland nog steeds te vuil: het milieu toen, nu en straks*, Houtekiet, Antwerpen.

Seattle Planning Department (1994) T*he Mayor's Proposed Comprehensive Plan; Toward a Sustainable Seattle*, Seattle Planning Department, Seattle, WA.

Seneca, L. A., C. Verhoeven (vertaler) (2004) *Brieven aan Lucilius*. Ambo, Amsterdam.

SenterNovem (2007) Instrumentenpalet duurzaam bouwen, bezocht: oktober 2007, http://duurzaambouwen.senternovem.nl/praktijk/gemeenten/instrumentenpalet_duurzaam_bouwen.

Smit, C.Th. (1989) Wet verontreiniging oppervlaktewateren, in: Brussaard, W., G.H. Addink (red.) *Milieurecht*, W.E.J. Tjeenk Willink, Zwolle.

Spit, T., P. Zoete (2002) *Gepland Nederland; Een inleiding in de ruimtelijke ordening en planologie*, Sdu Uitgevers, Den Haag.

Stadsgewest Haaglanden (2003) *Mirup; Handreiking 'Milieu in ruimtelijke plannen'*, Stadsgewest Haaglanden, Den Haag.

Stofberg, F.E., J.D.M. van Hal (1996) *Bouwstenen voor een duurzame stedebouw; 75 aanbevelingen voor een milieukundig ontwerp*, BOOM, Milieukundig Onderzoek- en Ontwerp Buro, VNG Uitgeverij, Den Haag.

Stuurgroep IMZS-Drechtsteden (1991) Rapportage 1e fase IMZS-Drechtsteden: inventarisatie milieubelasting, Provincie Zuid-Holland, Den Haag.

Tatenhove, J.P.M. van, H.J.M. Goverde (2000), Institutionalisering van milieubeleid, in: Driessen, P.P.J. en P. Glasbergen (red.), *Milieu, samenleving en beleid*, Elsevier, Den Haag.

Timár, E. (2005) Improving Environmental Performance of Local Land Use Plans; An Experiment with Sustainable Urban Planning in Amsterdam, in: Miller, D., G. de Roo (eds). *Urban Environmental Planning; Policies, Instruments and Methods in an International Perspective*, Ashgate, Aldershot, pp. 109–20.

Timmermans, W., J. Jonkhof, S. Tjallingii (2002) Strategie van de Twee Netwerken, in: Roggema,. R.E., A.J.M. Schreuders, M.O. Janssen, H. Hofstra, E.J. Vuijk, C.W. Jansen, W. Timmermans (red.) *Handboek Ruimtelijke Ordening en Milieu; Editie 2001–2002*, Kluwer, Alphen aan den Rijn, pp. 219–32.

Tjallingii, S. (1994) An Ecological Approach to Urban Planning; Strategies and Guiding Models, in: Vegt, H. van der, H. ter Heide, S. Tjallingii, D. van Alphen (red.) *Sustainable Urban Development; Research and Experiments*, Delft University Press, Delft, pp. 17–52.

Tjallingii, S. (1995) *Ecopolis; Strategies for ecologically sound urban development*, Backhuys Publishers, Leiden.

TK (Tweede Kamer) (1988a) Vierde Nota over de Ruimtelijke Ordening; Op weg naar 2015, deel A: beleidsvoornemers, Tweede Kamer, 1988–1989, 20490, nrs. 1–2, Den Haag.

TK (Tweede Kamer) (1988b) Vierde Nota over de Ruimtelijke Ordening; Op weg naar 2015, deel D: regeringsbeslissing, Tweede Kamer, 1988–1989, 20490, nrs. 9–10, Den Haag.

TK (Tweede Kamer) (1989) Nationaal Milieubeleidsplan; Kiezen of verliezen, 1988–1989, 21137, nrs. 1-2, Den Haag.

TK (Tweede Kamer) (1990) Actieplan Gebiedsgericht Milieubeleid, vergaderjaar 1990–1991, 21896, nrs. 1–2, Sdu Uitgevers, Den Haag.

TK (Tweede Kamer) (1993) Nationaal Milieubeleidsplan 2; Milieu als Maatstaf, 23560, nrs. 1–2, Sdu Uitgevers, Den Haag.

TK (Tweede Kamer) (2002) Nieuwe regels omtrent de ruimtelijke ordening (Wet ruimtelijke ordening), memorie van toelichting, 28916, nr. 3, Den Haag.

Van Ravesteyn, N. & D. Evers (2004) *Unseen Europe: A Survey of EU politics and its impacts on spatial development in the Netherlands,* Netherlands Institute for Spatial Planning, The Hague.

Van Tatenhove, J. B. Arts, and P. Leroy (red.) (2000). 'Political Modernisation and the Environment: The Renewal of Environmental Policy Arrangements. Dordrecht/Boston/London: Kluwer Academic Publishers.

Visser, J., C. Zuidema (2007) *De milieuatlas; Het milieu in contouren en mozaïeken*, Sdu Uitgevers, Den Haag.

VM (ministerie van Volkshuisvesting en Milieuhygiëne) (1972) Urgentienota Milieuhygiëne, Tweede Kamer, 1971–1972, 11906, nr. 2, Den Haag.

VM (ministerie van Volkshuisvesting en Milieuhygiëne) (1976) Nota milieuhygiënische normen 1976, Tweede Kamer, 1976–1977, 14318, nr. 2, Den Haag.

VNG (Vereniging van Nederlandse Gemeenten) (1986) Bedrijven en milieuzonering, groene reeks nr. 80, VNG-uitgeverij, Den Haag.

VNG (Vereniging van Nederlandse Gemeenten) (1999) Bedrijven en milieuzonering, milieureeks (kgm), VNG-uitgeverij, Den Haag.

VNG, VROM, UvW, IPO (Vereniging van Nederlandse Gemeenten; ministerie van Volkshuisvesting, Ruimtelijke Ordening en Milieubeheer; Unie van Waterschappen, Interprovinciaal Overlegorgaan) (2004) handreiking Milieukwaliteit in de leefomgeving; Werken aan gebiedsgericht maatwerk, VNG-uitgeverij, Den Haag.

Voerknecht, H.C. (1993) Casus De Drechtsteden; Saneren en bestemmen in een regionale aanpak, in: G. de Roo (red.), *Kwaliteit van norm en zone; Planologische consequenties van (integrale) milieuzonering*, Geo Pers, Groningen, pp. 83–95.

Voogd, H. (1999) *Facetten van de planologie*, Samsom Uitgeverij, Alphen aan den Rijn.

VROM (ministerie van Volkshuisvesting, Ruimtelijke Ordening en Milieubeheer) (1983) Plan Integratie Milieubeleid, Tweede Kamer, 1982–1983, 17931, nr. 6, Den Haag.

VROM (ministerie van Volkshuisvesting, Ruimtelijke Ordening en Milieubeheer) (1984a) Meer dan de som der delen; Eerste nota over de planning van het milieubeleid, Tweede Kamer, vergaderjaar 1983–1984, 18292, nr. 2, Den Haag.

VROM (ministerie van Volkshuisvesting, Ruimtelijke Ordening en Milieubeheer) (1989) Nationaal Milieubeleidsplan; Kiezen of verliezen, Ministerie van VROM, Den Haag.

VROM (ministerie van Volkshuisvesting, Ruimtelijke Ordening en Milieubeheer) (1990) Ministeriële handreiking voor een voorlopige systematiek voor de integrale milieuzonering, IMZ-reeks deel 6, VROM, Den Haag.

VROM (ministerie van Volkshuisvesting, Ruimtelijke Ordening en Milieubeheer) (1993) Nationaal Milieubeleidsplan 2; Milieu als maatstaf, Ministerie van VROM, Den Haag.

VROM (ministerie van Volkshuisvesting, Ruimtelijke Ordening en Milieubeheer) (1995) Waar velen willen zijn, is ook een weg; Stad & Milieu rapportage, Directoraat Generaal Milieubeheer, VROM, Den Haag.

VROM (ministerie van Volkshuisvesting, Ruimtelijke Ordening en Milieubeheer) (2001) Een wereld en een wil: werken aan duurzaamheid; Vierde Nationaal Milieubeleidsplan, VROM, Den Haag.

VROM (ministerie van Volkshuisvesting, Ruimtelijke Ordening en Milieubeheer) (2002a) Kennisboek Milieu in stedelijke vernieuwing; Een kwestie van inhoud en proces, VROM, Den Haag.

VROM (ministerie van Volkshuisvesting, Ruimtelijke Ordening en Milieubeheer) (2002b) Nader rapport inzake het voorstel van wet, houdende nieuwe regels omtrent de ruimtelijke ordening (Wet ruimtelijke ordening), Den Haag.

VROM (ministerie van Volkshuisvesting, Ruimtelijke Ordening en Milieubeheer) (2003a), Onderzoek Tweede Tussenevaluatie Stad & Milieu - Deel 1: relatie tussen norm en kwaliteit, VROM, Den Haag.

VROM (ministerie van Volkshuisvesting, Ruimtelijke Ordening en Milieubeheer) (2003b) Naar een optimale leefkwaliteit; Methoden voor milieukwaliteit in lokale plannen, VROM, Den Haag.

VROM (ministerie van Volkshuisvesting, Ruimtelijke Ordening en Milieubeheer) (2003c) Wetsvoorstel nieuwe Wet op de ruimtelijke ordening; Memorie van toelichting, VROM, Den Haag.

VROM (ministerie van Volkshuisvesting, Ruimtelijke Ordening en Milieubeheer) (2004a) Strategische Milieubeoordeling; Aandachtspunten inzake de toepassing van de Europese richtlijn 2001/42/EG voor Strategische Milieubeoordeling, VROM, Den Haag.

VROM (ministerie van Volkshuisvesting, Ruimtelijke Ordening en Milieubeheer) (2004b) Meer dan één stap vooruit!; Ervaringen uit 25 projecten Stad & Milieu, VROM, Den Haag.

VROM (ministerie van Volkshuisvesting, Ruimtelijke Ordening en Milieubeheer) (2004c) Milieu-informatie en openbaarheid, brief van staatssecretaris Van Geel aan gemeenten, september 2004, VROM, Den Haag.

VROM (ministerie van Volkshuisvesting, Ruimtelijke Ordening en Milieubeheer) (2006a) Actieprogramma Gezondheid en Milieu; Eindrapportage, VROM, Den Haag.

VROM (ministerie van Volkshuisvesting, Ruimtelijke Ordening en Milieubeheer) (2006b) Toekomstagenda Milieu; Schoon, slim, sterk, VROM, Den Haag.

VROM, LNV, VW, EZ (ministeries van Volkshuisvesting, Ruimtelijke Ordening en Milieubeheer; Landbouw, Natuurbeheer en Voedselkwaliteit; Verkeer en Waterstaat; Economische Zaken) (1997) Nota Milieu en Economie; Op weg naar een duurzame economie, VROM, Den Haag.

VROM, LNV, VW, EZ (ministeries van Volkshuisvesting, Ruimtelijke Ordening en Milieubeheer; Landbouw, Natuur en Voedselkwaliteit; Verkeer en Waterstaat; Economische Zaken) (2004) Nota Ruimte; Ruimte voor ontwikkeling, VROM, Den Haag.

VROM-raad (2005) Milieu en de kunst van het goede leven; Advies voor de Toekomstagenda Milieu, advies nr. 48, VROM-raad, Den Haag.

VWRR (Voorlopige Wetenschappelijke Raad voor het Regeringsbeleid) (1975) De organisatie van het openbaar bestuur; Enkele aspecten, knelpunten en voorstellen, Staatsuitgeverij, Den Haag.

WCED (World Commission on Environment and Development) (1987) *Our Common Future*, Oxford University Press, Oxford.

Weerd, P. de (2004) Deventer in kaart, in *VROM.nl*, jaargang 6, nr. 1, pp. 21–22

Index